AID FOR TRADE: GLOBAL AND REGIONAL PERSPECTIVES

United Nations University Series on Regionalism
Series editors: Philippe De Lombaerde and Luk Van Langenhove

The United Nations University Series on Regionalism, launched by UNU-CRIS and Springer, offers a platform for innovative work on (supra-national) regionalism from a global and inter-disciplinary perspective. It includes the policy-oriented *World Reports on Regional Integration,* published in collaboration with other UN agencies, but it is also open for theoretical, methodological and empirical contributions from academics and policy-makers worldwide.

Philippe De Lombaerde
Editor
Multilateralism, Regionalism and Bilateralism in Trade and Investment
1st World Report on Regional Integration

Philippe De Lombaerde and Lakshmi Puri
Editors
Aid for Trade. Global and Regional Perspectives
2nd World Report on Regional Integration

United Nations University

CRIS

Comparative Regional Integration Studies

Garnet

Aid for Trade: Global and Regional Perspectives

2nd World Report on Regional Integration

Philippe De Lombaerde and Lakshmi Puri (editors)

In Collaboration with:
UNCTAD, ECA, ECLAC, ESCAP, ESCWA, UNECE, UNIDO, UNDP, UNEP

Springer

Editors
Philippe De Lombaerde
United Nations University-
Comparative Regional
Integration
Studies (UNU-CRIS)
Potterierei 72
8000 Brugge
Belgium
pdelombaerde@cris.unu.be

Lakshmi Puri
UNCTAD
Palais des Nations
Av. de la Paix 8-14
1211 Geneve
Switzerland
Lakshmi.Puri@unctad.org

ISBN: 978-1-4020-9454-5 e-ISBN: 978-1-4020-9455-2

DOI: 10.1007/978-1-4020-9455-2

Library of Congress Control Number: 2008942062

Printed on acid-free paper

9 8 7 6 5 4 3 2 1

springer.com

ENDORSEMENTS

Aid for Trade (AfT) has become a major item on the international trade and development discourse. This is to a large extent in response to concerns expressed by developing countries and economies in transition with regard to their capacities to implement trade agreements, especially WTO agreements, and undertake necessary adjustments to increase net development gains from emerging trade opportunities.
In this ***World Report****, major UN agencies active in development cooperation and longstanding providers of trade-related technical assistance and capacity building discuss ways to sustain the momentum towards the operationalization and implementation of the AfT initiative and the supportive role to be played by the UN system. This is consistent with UN's role in promoting development and helping to achieve poverty reduction, as committed in the Millennium Declaration and the 2005 World Summit Outcome.*
The Report should be of particular interest to government officials, representatives of the private sector dealing with trade agreements and negotiations, civil society and academia.
Supachai Panitchpakdi, Secretary-General of UNCTAD.

Global governance is a concern of not only global organisations but also of all individual states and of regional co-operation agreements among sovereign states. In 2001, the United Nations University has launched UNU-CRIS, a research and training programme on comparative regional integration to study the role of regional integration in global governance. The ***World Report on Regional Integration*** *is a timely product of the research undertaken at UNU-CRIS. This new series of World Reports will certainly contribute to the discussion on the interaction between regional and global governance. And by bringing together insights from different parts of the UN system, in particular the five UN regional economic commissions and UNCTAD, this report will also contribute to a better understanding of the role of regions in the UN.*
Hans van Ginkel, former Under-Secretary-General of the United Nations and Rector of UNU.

The African, Caribbean and Pacific Group of States (ACP Group) welcomes the ***World Report on Regional Integration****, which provides a holistic and pragmatic perspective on the role of regions in world governance and international affairs. The most significant and unique value of this series is that it presents an enhanced understanding of the processes of regional integration in political, economic, social and cultural spheres. It is an invaluable reference for the ACP Group in its endeavour*

to further enhance solidarity and unity amongst its member states, whilst addressing the challenges of poverty eradication, sustainable development and globalization.
Sir John R. Kaputin, KBE, CMG, Secretary-General of the ACP Group.

The EU Framework VI GARNET Network of Excellence was created to bundle the capacity of more than 40 universities in order to jointly undertake innovative research on globalisation, regionalisation and the role of the EU, and regions in general, in world governance. In this context, GARNET actively supported the launching of this new ***World Report on Regional Integration****. The series is likely to establish itself as an obligatory reference for policy-makers and academics involved in thinking and moulding the global governance architecture of the future. It fills a gap in the existing supply of world reports.*
Professor Richard Higgott, Pro-Vice-Chancellor at the University of Warwick, and Senior Scientist and Director of the GARNET Network of Excellence.

TABLE OF CONTENTS

ACKNOWLEDGEMENTS

This Report, as the previous one, is the result of a collaborative effort involving several institutions and individuals. We want to mention the following people without whom this Report would not have been possible:

At UNCTAD: Secretary-General Supachai Panitchpakdi, Bonapas Onguglo, Lucian Cernat and Rajan Dhanjee.

At UNU-CRIS: Director Luk Van Langenhove, Birger Fühne, and Lurong Chen.

From the UN Commissions: Osvaldo Rosales and Mikio Kuwayama (UN-ECLAC); Hakim Ben Hammouda, Robert Okello and Joseph Atta-Mensah (UN-ECA), Nazem Abdalla and Abulgasim Abdullah, Mohamed Radwan and Jain Pak (UN-ESCWA); Ravi Ratnayake, Tiziana Bonapace, Mia Mikic, Marc Proksch and Noordin Azhari (UN-ESCAP); Virginia Cram-Martos and Lorenza Jachia (UN-ECE).

At UNDP: David Luke, Luca Monge-Roffarello and Sabrina Varma.

At UNIDO: François d' Adesky, Christophe Yvetot, Mohamed Dhaoui and Farrukh Alimdjanov.

At UNEP: Asad Naqvi.

At Boston University: Adil Najam.

From the GARNET Network of Excellence: Michael Schulz of Gothenburg University and Richard Higgott of Warwick University.

At Springer: Myriam Poort and Bernadette Ohmer.

We also thank Sylvie Capelle, Pascale Vantorre and Liesbeth Martens (all at UNU-CRIS) for their help in formatting the manuscript.

Lakshmi Puri, Geneva
Philippe De Lombaerde, Bruges

PHILIPPE DE LOMBAERDE AND LAKSHMI PURI

INTRODUCTION

In the aftermath of the formation of the WTO, many developing countries have expressed their concern over their capacities to implement the agreements and to undertake the necessary policy actions and adjustments to benefit from emerging trade opportunities. The transformation of that concern into the Aid for Trade initiative became salient during the course of the Doha round of multilateral trade negotiations over adjustment and competitiveness challenges facing developing countries, arising from, *inter alia*, preference erosion, textile quota elimination, implementing existing and future WTO Agreements, as well as their inherent lack of supply capacity and trade-related infrastructure deficits and deficiencies. Subsequently, the Sixth WTO Ministerial Declaration endorsed the concept of Aid for Trade, mandated the establishment of a Task Force to operationalize the concept, and mandated the Director General of the WTO to consult on financing for the initiative. The WTO Task Force was composed, met and issued a report with recommendations that was endorsed by WTO members. The WTO Director General continues to consult donors on financing.

Various initiatives by UN agencies including UNCTAD and regional commissions have been implemented to sustain the momentum towards the operationalization and implementation of the Aid for Trade initiative. The UN system is supportive of the Aid for Trade (AfT) initiative and its developmental goals, as committed in the 2005 World Summit Outcome. To this end, to be instrumental, the UN system needs to promote a more coherent approach and support towards the implementation of the AfT initiative.

This second volume of the World Report on Regional Integration wants to contribute to this aim by offering a UN-wide coherent perspective and a comprehensive contribution to the international policy debate on how to practically operationalize the AfT initiative in the different world regions and ensure that it contributes most effectively to the development dimension of the WTO negotiations. Global perspectives as well as regional and national perspectives on the challenges and opportunities in demanding and supplying aid for trade, consistent with the recommendations of the WTO Task Force, are presented.

The questions that are addressed in this World Report include the following:

(1) What are the specific AfT-related needs and priorities in each developing region and from a cross-regional perspective, using as a basis the scope and areas of focus identified by the WTO Task force? Apart from broad regional perspectives, what are the needs of specific countries (LDCs, small vulnerable states,

P. De Lombaerde, L. Puri (eds.), *Aid for Trade: Global and Regional Perspectives*,
United Nations University Series on Regionalism 2,
DOI 10.1007/978-1-4020-9455-2_1, © Springer Science+Business Media B.V. 2009

landlocked economies, etc)? What are the needs of countries with economies in transition?
(2) What role can global and regional organizations and processes play to help developing countries identify needs, successfully prepare projects and mobilize donor support?
(3) Which best suited institutional mechanism at national, regional and global levels need to be set up for planning, strategizing, conceptualizing and implementing AfT? Within this process, what is the role of UN agencies in the emerging AfT architecture in partnership with other organizations – intergovernmental, public, private and NGOs?

The World Report is organized in two parts and a Statistical Annex. The first part offers a global perspective on AfT, highlights the contributions of the UN system to AfT, and is complemented with perspectives from think tanks. Chapters are written by authors from UNCTAD, UNDP, UNEP, UNIDO, UNECE, UNU-CRIS, GDN and ECDPM.

The second part offers regional perspectives and looks at the potential for AfT from the angle of the specific characteristics and needs in the different regions in the world. Chapters are written by the UN regional commissions (ECA, ECLAC, ESCAP, ESCWA).

A number of concrete proposals and policy recommendations are included in the different chapters.

The views of the UN regional commissions have also been presented to the policy community at the Aid for Trade parallel event at the UNCTAD XIIth Ministerial Conference, held in Accra (Ghana) from 20–25 April 2008 (www.unctadxii.org).

As in the previous Report, the volume concludes with some key data on regional organizations in the Statistical Annex. These data include: memberships, year of establishment, links to official websites, shares of world population and GDP, and intra-regional trade intensities.

This World Report and the statistical information it contains, have only academic and informative purposes. They do not, in any way, imply official statements of the United Nations, the United Nations University or any other UN organ regarding regional organizations or UN member states. The authors are personally responsible for the contents of their chapters.

The series of World Reports was launched in 2007. The first volume was published as *Multilateralism, Regionalism and Bilateralism in Trade and Investment. 2006 World Report on Regional Integration*, also by Springer (www.springer.com).

PART I

AID FOR TRADE: GLOBAL PERSPECTIVES

UNCTAD

CHAPTER 1

AID FOR TRADE AND DEVELOPMENT

1.1. INTRODUCTION

For several decades 'trade, not aid' was a prevalent dictum. In the mid-1960s, when the founding father and first Secretary-General of UNCTAD, Raul Prebisch (1964) coined the 'trade gap' concept, it became apparent that official development assistance (ODA) would be insufficient to provide the necessary foreign exchange resources that developing countries would need to import the capital goods that are so crucial for enhancing productive capacity, trading and moving up the development ladder into more value-added and higher-paying economic activities. The 'trade, not aid' philosophy was also justified by a number of imperfections of aid and aid policies.

Over the last few years, the difficulties that the vast majority of developing countries have faced in benefiting from economic and trade reforms and promoting development have revealed that the 'trade, not aid' and even 'trade as aid' prescriptions also have their limits. Today, the logic has evolved from 'trade, not aid' and 'trade and aid' to including 'aid for trade'. This conceptual progression is important in that it marks a significant step forward by the international community in accepting that trade-specific development assistance must accompany any trade reform effort for such efforts to be meaningful and lasting in terms of their development impact. Understanding the logic between aid, trade and development has posed enduring challenges to development economists. The aid literature is very voluminous and multifaceted.[1]

By its very nature and mandates, the United Nations system plays an important role in aid for trade, including through the provision of trade-related technical assistance, as an integral aspect of its efforts to promote development and help achieve poverty reduction. This includes mobilizing global partnerships for development to meeting internationally agreed development goals, including those set out in the Millennium Declaration[2] and the 2005 World Summit Outcome.[3]

As the focal point of the United Nations system on the integrated treatment of trade and development, UNCTAD has contributed to the conceptualisation of the aid for trade initiative. As part of the development component of the United Nations system, UNCTAD has since its creation in 1964 provided trade-related and capacity-building support – i.e. aid for trade – to developing countries and countries with economies in transition to effectively integrate into and realize development benefits from the international trading system. This paper identifies UNCTAD's contribution to aid for trade and its role in support of the WTO in advancing the implementation of

P. De Lombaerde, L. Puri (eds.), *Aid for Trade: Global and Regional Perspectives*,
United Nations University Series on Regionalism 2,
DOI 10.1007/978-1-4020-9455-2_2, © Springer Science+Business Media B.V. 2009

the aid for trade initiative for the benefit of developing countries and countries with economies in transition.

1.2. AID AND TRADE

More development assistance has become a clarion call of today's development dialogue. Once again, there are calls for an increase in ODA, in line with various internationally agreed targets,[4] coupled with forms of other international development support such as favourable market access.

This turnaround in the international consensus on the trade, development and financial nexus follows from earlier dissatisfaction with the focus on development assistance owing to a number of imperfections with aid and aid policies. The overall record of aid for development policies shows results that are disappointing or mixed and often complicated by institutional factors.[5] Some earlier analyses were concerned primarily with the importance of 'tied aid' effects on increased imports of services or goods from the donor.[6] Other studies concluded that the causality is not unidirectional and both aid and trade flows and policies can have a mutual influence. Aid for trade may increase trade flows through the general economic effects it induces in the recipient country. Aid for trade flows may also increase exports directly when its effects are tailored to export promotion.[7] Others argue that aid flows may have a distinct import-augmenting effect in the recipient country, especially when such flows are tied to goods and services procured from donor countries (Kemp and Kojima, 1985).[8]

Sceptical opinions about aid were common among experts and academics as well as among some donor countries and beneficiary developing countries. Aid per se or the mechanisms through which it is disbursed can lead to detrimental developmental effects or unintended disincentives in beneficiary countries. Some critics pointed to the danger of the 'aid curse', where the potential negative effects would prevail and make matters worse rather than better for development, similar to the 'resource curse' or 'Dutch disease' effect.[9] Page (2006), for instance, identified several potential conflicts between aid and trade: high levels of aid can lead to 'Dutch disease' and result in undermining trade, thus defeating the very purpose of aid for trade (Page, 2006). This is especially true when the absorptive capacity of the beneficiary country is weak and when aid flows are considerable relative to GDP, overall foreign exchange reserves or Government spending. Consequently, aid alone is a necessary but not a sustainable solution to escaping the poverty trap and meeting other development challenges.

Conversely, several success stories are also available to show that under the right conditions, aid may provide a strong impetus to a thriving development strategy. The early examples that are usually given by aid proponents are Japan, Republic of Korea and Taiwan Province of China, between 1950 and 1970s. More recently, a few other developing countries (particularly Asian nations) have showed that external development aid can play a positive role as part of a well-articulated and sustained development strategy.

Over the last few years, the difficulties which the vast majority of developing countries have experienced in harnessing trade liberalisation to integrate more effectively and beneficially into international trade flows and promote development have revealed that the 'trade, not aid' and even 'trade as aid' prescriptions also have their limits (Hoekman and Prowse, 2005). This insight comes at an occasion even after very ambitious trade opening schemes were put in place in favour of developing countries and especially the least developed countries (LDCs). These include such schemes as the EU's Everything But Arms initiative (EBA), the United States' African Growth Opportunity Act (AGOA) and similar initiatives in other OECD countries, including under the Generalised System of Preferences (GSP). The difficulties which beneficiary countries of these schemes faced in utilizing the schemes highlighted factors hindering the effective implementation of such schemes, both endogenous (such as rules of origin) and exogenous (such as supply capacities and competitiveness).

Moreover, the adoption of the WTO Agreements in 1995 with a very comprehensive trade liberalisation agenda brought into sharp focus the difficulties many developing countries faced in implementing these Agreements. Many of the agreements were very complex and costly to put in place, as they required cumbersome administrative structures and major legal and regulatory changes. Such concerns led to the emphasis on, inter alia, 'implementation-related issues and concerns' in the Doha Round of multilateral trade negotiations.

Hence, there has emerged an international consensus that aid for trade must be provided in tandem with any trade reform agenda for the latter to be successful and meaningful in terms of its development impact. For example, the report of the United Nations Millennium Project[10] suggested among other things the creation of an incremental and temporary 'aid for trade fund' commensurate with the magnitude of the task of helping countries cope with adjustment costs associated with the implementation of the outcome of the Doha Round of multilateral trade negotiations. The Sixth WTO Ministerial Conference agreed that "Aid for Trade should aim to help developing countries, particularly LDCs, to build the supply-side capacity and trade-related infrastructure that they need to assist them to implement and benefit from WTO Agreements and more broadly to expand their trade."[11] The WTO process has come to be labelled as the '*Aid for Trade Initiative*'.

While empirical literature on aid for trade does not provide robust results so far regarding the impact of aid and trade flows, it does reveal a positive correlation. For instance, Hoekman and Olarreaga (2007) conclude by pointing out the imperative need for complementary policies, including aid for trade, for trade reform to have a significant contribution to poverty reduction efforts. UNCTAD refers to such complementary policies as the '*enabling environment*'. Without substantial aid for trade to help ensure that trade reforms are implemented and weaker economies are equipped to handle international competition so that the expected benefits accrue to developing countries in particular, the poverty impact of, for example, the Doha Round outcome is likely to be marginal at most. This raises the question of the rationale of trade liberalisation and the resulting trade agreements if, after many years of negotiations, poor people in poor countries living on less than one dollar a day would only gain a few more cents.[12] In this connection, aid for trade should be used to ensure that no nation is made worse off by the outcome of the Doha Round (Evenett, 2005). Such an approach would call for an ambitious aid for trade initiative, which should be functioning prior to the expected liberalisation. This assistance could serve as the transmission belt between trade liberalisation, successful development and poverty alleviation for developing countries.

1.3. AID FOR TRADE

Aid for trade in connection with adjustment support to trade liberalisation has been an accepted practice at the national and regional level for many years, especially in developed countries, where governments have put in place financial compensation and other schemes to help sectors and workers that are likely to be adversely affected by such reforms. The generalisation of this concept at the international level, however, was lacking until the Aid for Trade Initiative was revived. There is now renewed interest and political commitment in favour of 'aid for trade' as a necessary complement to trade reform and thus a necessary complement to sustainable development. At the same time, there is also general recognition that aid for trade is not a substitute or precondition for trade liberalisation, but rather a valuable complement thereto.

Following the Sixth WTO Ministerial Conference, international consideration of the Aid for Trade Initiative has been largely focused on defining the potential policy areas for such assistance; the modalities through which such an initiative can be implemented; and its regular review and monitoring. Pursuant to the mandate from the Sixth WTO Ministerial Conference, a WTO Task Force on Aid for Trade was set up to address the ways in which aid for trade might most effectively contribute to the development dimension of the Doha Round. One of the main concerns in the early stages of the aid for trade discussions pertained to definitional aspects: what is aid for trade, what are the ongoing efforts and what else can be included in this definition (OECD, 2006)? Another area of work was concerned with the institutional mechanisms of aid for trade delivery (Stiglitz and Charlton, 2006). Many of these conceptual issues were discussed and highlighted in an international conference on aid for trade convened by UNCTAD and the Commonwealth Secretariat in March 2006 (UNCTAD, 2006). Drawing on these assessments and other inputs, the WTO Aid for Trade Task Force, inter alia, identified six policy areas for aid for trade assistance (WTO, 2006):

(1) Trade policy and regulations,
(2) Trade development,
(3) Trade-related infrastructure,
(4) Building productive capacity,
(5) Trade-related adjustment, and
(6) Other trade-related needs.

This delineation of aid for trade by the WTO Task Force shows that aid for trade is not an entirely new concept, especially when viewed from the traditional perspective of trade-related technical assistance and capacity-building. Table 1.1 shows the volume of aid for trade that is being provided. These aid for trade funds are disbursed through various projects and operational activities, all designed to tackle serious trade-related challenges faced by developing countries. From 2001 to 2005, $53.7 billion was allocated by developed countries to developing countries under various trade-related assistance programmes covering three broad categories, namely trade policy-related capacity-building; trade infrastructure; and trade development.

From a regional perspective, Asia accounts for more than 40 per cent of total aid for trade funds, followed by Africa (28.8 per cent) and Latin America (12.8 per cent). Among the broad categories of aid for trade, trade development represents more than half of total funds (56 per cent), trade infrastructure around 41 per cent and trade policy capacity-building represents a small proportion of current funds (3 per cent).

Table 1.1 also shows that the structure of aid for trade across regions varies considerably. For instance, almost 80 per cent of aid for trade to Latin America was dedicated to trade development and 19 per cent to trade infrastructure. In North Africa and Middle East, 20 per cent of aid for trade has been geared to trade policy capacity-building, compared to 1–3 per cent for such aid for trade categories across all other regions, a trend which points to the diversity in the utilisation of aid for trade across developing regions. This diversity is even greater within regions in terms of aid for trade data at the country level. Further research is needed on whether or to what extent this diversity in aid for trade utilisation is related to differences in the needs or requests of aid for trade recipients or rather reflects variances in donor priorities.

Thus, aid for trade and development includes: assistance to build human, institutional and regulatory capacities in developing countries to formulate and implement locally owned trade policies; participate in and shape the outcome of trade negotiations; implement trade agreements; build supply-side capacities (including trade-related infrastructure) to take advantage of new trading opportunities; and provide compensatory assistance to offset adjustment costs in sectors adversely affected. It also includes assistance to help developing countries deal with the increase in anti-competitive practices which may occur as trade liberalisation exposes enterprises to more competition; and improve the competitiveness of commodity sectors. These could constitute possible specific areas of trade policy in any aid for trade initiative for development.

Trade policy support is important in terms of enabling many developing countries, particularly those with limited administrative and financial capacities, to participate effectively and project their commercial and development interests in trade negotiations and norm-setting processes at multilateral and/or regional levels. Many developing countries, especially LDCs, lack well-trained trade officials and negotiating apparatuses that could reflect the needs of their private sector and specific economic sectors in clear negotiating positions. Nor do they have the capacity to leverage existing multilateral trade rules into trading opportunities for their exporters. Such constraints can be addressed by aid for trade.

TABLE 1.1. Aid for Trade, by region and type of aid (2001–2005) ($ million)

	Trade policy	Trade infrastructure	Trade development	Total aid for trade			
	$ million	%	$ million	%	$ million	%	$ million
Sub-Saharan Africa	292	2	6,188	40	9,001	58	15,481
Asia	213	1	10,946	50	10,685	49	21,845
Latin America	176	3	1,295	19	5,418	79	6,889
North Africa and Middle East	809	20	1,343	33	1,976	48	4,128
Transition economies and other Europe	163	3	2,243	41	3,013	56	5,418
Total	1,652	3	22,014	41	30,094	56	53,760

Source: OECD Creditor Reporting System database and UNCTAD calculations.
Notes: The aid for trade considered in this analysis comprises only those flows disbursed, in contrast to the aid committed. As a result of this methodology, the total aid for trade figures reported may be different from other estimates of aid for trade in the literature.

Assistance on *trade-related adjustment, including support to developing countries in putting in place accompanying measures that assist them to benefit from liberalised trade*, is another priority area. There is the important and more immediate need for developing countries to cope with trade shocks and costs associated with liberalisation and the difficult transition to a more liberalised global trading environment. Aid for trade can help ensure that, alongside trade liberalisation, there is also adjustment towards growth sectors which can help obviate or minimize serious costs to especially poor people. This is particularly important in most developing countries, as they lack their own social safety nets. The costs of such adjustment, either in budgetary terms of setting up institutional structures to meet trade obligations, or in terms of loss of jobs and possible deindustrialisation from closure of enterprises which are unable to compete, can be overwhelming for developing countries. They can nullify the potential benefits of market openings and act as a deterrent to liberalisation or to implementation of trading obligations. Adjustment support is particularly relevant for those countries that are dependent on trade preferences that are being gradually eroded with globalisation. Laird (2007) makes a convincing case in favour of the need to include adjustment costs arising from multilateral trade negotiations as part of the aid for trade initiative (Laird, 2007).

Trade facilitation, which plays a central role in determining the competitiveness or lack thereof of developing country exporters, is another major area for aid for trade support. It is intrinsically linked to *transport costs* which increase with deficient transport and trade facilitation infrastructure. In LDCs and landlocked countries, high transport costs and poor infrastructure have a critical impact since they can affect not only the volume of trade but also even the possibility for trade to take place at all. Trade costs estimates are often more significant than ad valorem tariffs. Francois and Manchin argue that transport infrastructure is even more important in explaining trade growth than tariff reductions (François and Manchin, 2007). It has therefore been suggested to extend some of the current aid for trade categories to other infrastructure and policy elements that have an impact on the export "value chain" (Wilska and von Bonsdorff, 2006).

In the medium to long term, the building-up of *productive capacity* is a sine qua non for the trade integration of developing countries, and must therefore be a central feature of aid for trade. This calls for using aid for trade as well as development assistance generally more effectively, to support proactive measures in areas such as commodity sector development, industrial development, services sector development and trade, and the associated supportive infrastructure, including physical, national financial and debt management systems, investment promotion (domestic and foreign), entrepreneurship and enterprise development, and trade-related infrastructure and logistics.

UNCTAD's *Trade and Development Report 2007*, for example, finds that regional trade liberalisation among developing countries can help promote and advance integration into the global economy. However, trade liberalisation is a necessary but insufficient condition for trade growth. Other factors, such as good transport connections may be equally or more important for trade growth and competitiveness. Thus, the various factors coming into play in the trade picture should be supported, including through aid for trade. In this connection, the creation of a $1 billion Aid-for-Trade Fund, which would be in addition to aid for development (including with regard to LDCs), has been suggested (Puri, 2005). The facility would provide much-needed finance to meet adjustment costs arising from trade reform, deal with trade-related infrastructure constraints, and upgrade supply capacity and competitiveness in LDCs.

As analysed in aid and trade literature, the complementarity between aid and trade has to be considered as a complex interaction, going beyond a simple capital transfer between donor and recipient countries. Aid for trade may create various positive externalities that go beyond the specific areas of intervention. Aid for trade may also lead to technology transfers, create networks among key trade-related institutions, and disseminate best practices. The complexity of the issues involved indicates that more in-depth country-level and regional-level analysis of aid for trade contribution and impact is needed. UNCTAD, WTO and other agencies actively involved in the current aid for trade debate could play a useful role in undertaking such an analysis.

1.4. UNCTAD'S CONTRIBUTION

1.4.1 UNCTAD's Mandate

By its very nature and mandates, the United Nations system has an important role to play in aid for trade through both existing and new established mechanisms. This is consistent with the UN's role in promoting development and helping to achieve poverty reduction, in line with the commitments set out in the Millennium Declaration and the associated Millennium Development Goals, as well as the 2005 World Summit Outcome. As early as 2003, the United Nations Secretary-General, in his address delivered at the Fifth WTO Ministerial Conference, highlighted the need for aid for trade, in addition to aid for development, as a necessary complement of market access negotiations in the context of wider international development cooperation efforts to promote timely achievement of internationally agreed development goals.[13]

UNCTAD's aid for trade strategy, as part of its trade-related technical assistance activities, is derived from its core functions as the United Nations focal point on the integrated treatment of trade and development, and from the development objectives set by member States at UNCTAD's quadrennial conferences, most recently under the Accra Declaration and Accra Accord, adopted by UNCTAD XI. UNCTAD approaches aid for trade through its three pillars of work, namely intergovernmental policy dialogue and consensus-building, research and analysis, and technical cooperation and capacity-building. The emphasis is on a holistic approach to integrating developing countries into the international trading system, both quantitatively and qualitatively.

As mentioned above, immediately after the WTO Ministerial Conference which launched the WTO Aid for Trade Initiative, UNCTAD, in collaboration with the

Commonwealth Secretariat, convened an international conference on aid for trade. The event, held in March 2006, provided insights on key aspects regarding the possible operationalisation of the aid for trade initiative. It also highlighted the private sector dimension as well as the regional dimension of aid for trade, in complementing national-level support as well as multilateral provision of global public goods.

Subsequently, UNCTAD has contributed to efforts by developing countries and their groups, as well as that of the WTO Task Force on Aid for Trade, to define aid for trade, its components and the modalities of its implementation. As part of the preparatory process for UNCTAD XII, UNCTAD coordinated with ESCAP an aid for trade event in January 2008 at which different United Nations agencies involved in aid for trade discussed their own perspectives and how their different contributions could be coordinated for enhanced impact.[14] This was followed up at UNCTAD XII itself, where a side event on "Aid for Trade: perspectives of Regional Commissions" was organized by UN regional commissions: each of them presented their perspectives on Aid for Trade (in particular, how they saw its implementation) and their activities in the field, following which an interactive discussion on the subject was held.

Such efforts have also informed UNCTAD's own vision and strategy regarding aid for trade. Significantly, following the mid-term review of the Sao Paulo Consensus adopted by UNCTAD XI in 2006, UNCTAD's member States agreed that UNCTAD should emphasize "Aid for Trade, including aid for institutional, regulatory, infrastructural and human resources development in developing countries;" that UNCTAD's technical cooperation pillar should be strengthened by "ensuring that UNCTAD is in a position to play an important role in the Aid for Trade Initiative, in accordance with UNCTAD's mandates, expertise and development approach;" and that UNCTAD should "assume an important role in providing assistance to developing countries under the Aid for Trade Initiative, taking into account their national development strategies."[15] UNCTAD XII is expected to consolidate further the framework of UNCTAD's work on aid for trade.

1.4.2 UNCTAD's Activities

UNCTAD's definition of aid for trade follows broadly the definition adopted by the WTO Task Force on Aid for Trade. It looks at aid for trade in terms of both the *soft* aspects of trade (trade and investment policies and negotiations, for example) and the *hard* aspects of trade infrastructure (for example, trade adjustment, building supply capacity and competitiveness, trade facilitation, transport connectivity). The key elements of UNCTAD's aid for trade activities consist of building human, institutional, regulatory, analytical and trade infrastructure capacities. Such activities also promote building of supply capacity and competitiveness, trade-related infrastructure-building (including through trade facilitation and enterprise development), and building negotiating capacities and institutions to secure effective market access and market entry.

Over the years, UNCTAD has been one of the main agencies engaged in trade-related technical assistance activities. It provides trade and trade-related assistance and capacity-building support at each stage of the trading process, from investment, enterprise development and financing, through customs operation and transport, to market access and market entry. UNCTAD lends its expertise to developing countries in areas such as the following:

- Participating effectively in trade and investment negotiations;
- Providing assistance on WTO accession;
- Diversifying commodity-dependent economies;
- Developing services sectors;
- Addressing non-tariff barriers and the interface between trade and environmental measures;
- Meeting product standards and regulations;
- Formulating national and regional trade and trade-related policies, and investment policies;
- Drafting and implementing competition law and policy;
- Assessing trade-related needs and impacts of trade agreements;
- Supporting adjustment to trade reform;
- Assisting with debt management;
- Building trade facilitation infrastructure such as customs automation;

UNCTAD's approach is that the aid for trade should be specific to trade capacity development, and thus additional to development assistance; benefit all developing countries that demand it; and meet national and regional needs as well as deliver global public goods that can be made available to all. Global public goods that can be provided by UNCTAD for the benefit of all developing countries include: UNCTAD's trade databases and analytical softwares such as a new database on non-tariff barriers; development of trade negotiation capacities; provision of assistance on WTO accession; activities to meet emerging product standards and environmental norms affecting trade; customs automation systems; debt management; training of trainers, policymakers and trade negotiators, including through linkages with universities and academic institutions; investment policy tools; and promotion of the Enhanced Integrated Framework of Trade Related Technical Assistance for LDCs.

As a 'think tank' with a long history of intellectual contributions on trade and development, UNCTAD also assists with the conceptualisation and design of country-specific and regional aid for trade programmes, and supports their implementation. UNCTAD has contributed to the conceptualisation of the Aid for Trade Initiative through studies and high-level multi-stakeholder meetings to clarify the concept and suggest appropriate ways and means to implement it.

UNCTAD's aid for trade activities are demand-driven and respond to the needs and priorities formulated by the beneficiary countries in their requests for technical cooperation. One important objective is the improvement of institutional and human capacity constraining the ability of many developing countries to undertake in-country trade policy formulation and prioritisation, and development of trade infrastructure.

UNCTAD has also addressed the lack of institutional and process linkages between trade policies and national development strategies, such as trade facilitation activities on transport connectivity. Whenever relevant, UNCTAD supports appropriate national needs assessment involving consultations with all relevant stakeholders. For example, under the Joint ITC/UNCTAD/WTO Integrated Technical Assistance Programme for African Countries (JITAP), UNCTAD mobilised the three agencies' expertise to build and strengthen in 16 African countries their inter-institutional coordination among all key stakeholders (Government, private sector and civil society) in a participatory approach to WTO trade negotiations and adjustment to and implementation of trade agreements. Such an inclusive approach is a prerequisite for successful aid for trade, as needs and priorities, by virtue of country ownership, would have to be ascertained by beneficiary countries themselves.

The scale and scope of UNCTAD's aid for trade activities enables it to provide customised trade-related assistance to developing countries and countries with economies in transition. Its accumulated experience and outreach enables it to have a multiplier effect across countries, thematic areas, and sectors. UNCTAD's comparative advantage in this area is its ability to provide beneficiary countries with an integrated perspective on trade and development strategies and policies, and to promote coherence in this regard at national, regional and international levels.

Aid for trade activities, as part of UNCTAD's overall technical cooperation and capacity-building activities, also draw on and benefit from intergovernmental policy dialogue and consensus-building, as well as sound research and policy analysis. This three-pronged approach is unique, and ensures that aid for trade activities are fully integrated into UNCTAD's development work.

Addressing development priorities at the regional level is one of the main priorities of UNCTAD's work. It has been providing a wide portfolio of aid for trade activities at regional level, such as promoting regional institutional development; building regional coordination mechanisms; assisting developing countries in regional harmonisation of technical regulations and standards; helping with trade negotiations in regional contexts (both South–South and North–South); enhancing transport and trade logistics; linking national/regional systems to global transport operators and networks; or providing assistance for trade facilitation. Such a regional strategy includes tailored support to a particular region, for example, Africa, Latin America or Asia; support to a particular subregion, for example, Central America; or a thematic focus linked to a particular region, such as helping institutions meet SPS standards or environmental norms or regional services negotiations (for example for SADC), or assisting institutions with the implementation of competition policies.

Mainstreaming trade into national development strategy is a major concern of UNCTAD particularly because, so far, only a minority of country-level plans include trade-related policies and assistance among their priorities. UNCTAD's advisory services emphasize the need for better integration of trade into development strategies that promote poverty reduction in developing countries, in particular in LDCs.

Although UNCTAD is a non-resident agency and has no established network for in-country presence, it is closely involved in the One UN pilot countries to promote joint United Nations support to a country and aims to integrate aid for trade into the consolidated United Nations country development programmes. UN-wide activities will also benefit from enhanced coherence and mutual support between the expertise of various United Nations agencies and their aid for trade activities and the trade-related components of national development plans. UNCTAD also participates in the United Nations Cluster on Trade and Productive Capacity, recently established by the Chief Executives Board for Coordination. In addition, it participates in the Trade Cluster of the United Nations Executive Committee on Economic and Social Affairs, under which aid for trade activities are coordinated among UNCTAD, UNEP, UNDP and United Nations regional commissions. All these efforts will ensure greater coherence among United Nations agencies on aid for trade and other development-oriented efforts, as well as bring about greater synergy between policy level and operational activities of these agencies, and further inform intergovernmental deliberations.

Such coordination is important as United Nations' aid for trade work is delivered through a variety of mechanisms (bilateral, multi-donor funded programmes, individual international organisations and agencies' programmes, multilateral agencies, regional organisations and regional financial institutions), with a multiplicity of programmes. Frequent fragmentation and lack of coordination among the different players remain major concerns, for both the agencies concerned and the beneficiaries. The present book, which constitutes a joint effort among UNCTAD, United Nations regional commissions, UNDP and UNEP as well as the United Nations University (Comparative Regional Integration Studies), is an excellent example of coordition among different agencies. The Trade and Development Board, at its forty-first executive session in April 2007, agreed that UNCTAD's technical cooperation activities should be adapted and consolidated into a few overarching thematic themes so as to maximize impact on development and increase efficiency and coherence. Thus, the elements of the consolidated thematic approach are being developed in line with this recommendation.

UNCTAD's role in this area was boosted by UNCTAD XII, which placed a strong emphasis on Aid for Trade. The Accra Declaration (para. 11), calls for stronger national action and international support to help build domestic productive competitive export supply capacities, as well as trade-supportive logistics for developing countries, including Africa, LDCs, landlocked developing countries, small island developing States and other structurally weak and vulnerable small economies and declares that Aid for Trade could make a meaningful contribution to respond to these challenges and should be adequately resourced and implemented, through multilateral and bilateral channels, with the United Nations and other multilateral organizations playing an important role. The Accra Accord (para. 61), emphasizes that: Aid for Trade is an important tool for international trade and development and can help all developing countries, especially LDCs, to meaningfully exploit market access opportunities; its implementation and effective use can help developing countries to put in place accompanying measures that assist them in implementing and benefiting from WTO agreements and more broadly in expanding their trade; and it cannot be a substitute for the development benefits resulting from a successful and balanced outcome of the Doha Round but will be a valuable complement to them. Para. 71 of the Accord provides that: Aid for Trade, including effective technical assistance, should aim to help developing countries in their efforts to implement and benefit from trade liberalization and reform so as to build productive capacities and trade-related infrastructure based on each country's needs and priorities; increased and more effective Aid for Trade is needed to support all developing countries, in particular LDCs, to benefit from the rule-based international trading system; additional, predictable, sustainable and effective financing is fundamental to fulfilling the Aid for Trade mandate; and beneficiary countries should mainstream trade and integrate Aid for Trade into their development strategies in accordance with their national priorities., as a basis for effective and sustainable support. Para. 106 provides that UNCTAD should continue to play an important role in developing and implementing projects under Aid for Trade and to provide focused support to developing countries, in building their capacity to meaningfully participate in, and benefit from, the international trading system and to address the opportunities and challenges that this system offers.

Data relating to UNCTAD's allocation of the aid for trade share in individual projects and programmes and its commitments by aid for trade category for the period 2002–2005 and in particular 2005 are provided in the tables in the Annex. In 2006, there was a decrease in UNCTAD's interregional expenditure, despite a 16 per cent increase in overall technical assistance delivery. Interregional projects from which all developing regions benefited accounted for almost 50 per cent of total expenditure, down from over 52 per cent in 2005. At the same time, the share of regional projects increased from 9.6 per cent in 2005 to 11 per cent in 2006. The share of country programmes remained constant at around 40 per cent during this period.

UNCTAD is not a donor agency and is therefore not in a position to make aid for trade pledges. Rather, it is an implementing agency depending to a large extent (over 90 per cent) on voluntary contribution from donors for its aid for trade and trade-related technical assistance activities. The three main sources of financing for UNCTAD's technical cooperation in 2006 are indicated in Box 1.1.

Box 1.1 The three main sources of financing for UNCTAD technical cooperation, 2006

Trust funds 91.4%	*UNDP 3.9%*	*United Nations programme budget 4.7%*
$32.2 million (an increase of 19.7% over 2005)	$1.4 million (a decrease of 26.3% over 2005)	$1.7 million (a decrease of 2.4% over 2005)

1.4.3 Some Lessons from UNCTAD's Experience

As one of the main agencies in the United Nations system involved in providing trade-related technical assistance in furtherance of development, UNCTAD plays an important role in the realisation of the Aid for Trade Initiative. Implementing the Initiative more fully will be a key element of global solidarity for development in coming years. UNCTAD stands ready to support this process, as it is already doing so, and help, in cooperation with WTO and other United Nations agencies, with the ongoing Aid for Trade Initiative. In the light of UNCTAD's experience in aid for trade and trade-related technical assistance, some lessons learnt are highlighted below on how to ensure that aid for trade maximizes gains from trade and trade liberalisation for developing countries.

The principles guiding the Aid for Trade Initiative (namely 'secure', 'additional', 'predictable' and 'non-debt creating funding') and the Paris Declaration principles should be mutually reinforcing. Secure and predictable funding should, to the extent possible, be accompanied by the provision of adequate levels of resources commensurate with the needs of developing countries. The overall level of aid for trade resources that will be made available must reflect the growing demands of developing countries, in particular LDCs, and their genuine trade-related needs. Such assistance needs to be trade-specific and should not be subsumed into wider adjustment programmes.

In addition, the principle of country ownership will be critical for the success of the Aid for Trade Initiative and in particular for aid effectiveness. Ownership would include assessment and articulation of country needs, inter alia priority-setting, and their integration into development plans and poverty reduction strategies and effective participation in the governance structure for the management of the Aid for Trade Initiative. Accessibility, user-friendliness and non-debt-creating aspects are vital to ensure the successful impact of the Aid for Trade Initiative.

Over the years, it has become evident that trade-related technical assistance requires collaboration and pooling of synergies of implementing partners and relevant international organisations, each in line with each one's comparative advantage, and the drawing of lessons from existing delivery instruments or frameworks. Such cooperation is required to avoid competition for limited donor funds and to minimize duplication of efforts and wastage of resources.

Building local institutional capacities is important as such institutions in most developing countries are weak, a fact that undermines their ability to effective formulate and implement trade policy, develop trade negotiations strategies, and absorb trade related technical assistance. For this reason, training activities remain essential to empower trade-related institutions in developing countries. Thus, UNCTAD places a high priority on human resources development in trade issues.

Increased efforts need to be devoted to the improvement of trade-related 'global public goods' that enhance the ability of developing countries to take advantage of the opportunities offered by the international trading system. The provision of such 'global public goods' requires considerable resources, inter-institutional cooperation and the establishment of a network of country-level focal points. Greater use of and partnering with local institutions and expertise in developing countries helps with building capacities, transferring knowledge and expertise; minimizes costs; and facilitates exchange of experience and networking between countries and their policymakers.

Country-specific needs assessment for aid for trade is crucial. These assessments must be underpinned by empirical research and analytical capacities, as well as feedback and endorsement received through UNCTAD's intergovernmental deliberations and related mandates. Specific aid for trade activities will have to respond to both immediate needs, such as trade negotiations, and to medium to long-term needs of capacity-building, such as supply capacity and adjustment support. Customised trade-related assistance (sectoral, national, regional) is most appropriate, in keeping with the respective development and poverty reduction priorities of developing countries.

To achieve efficient results, the aid for trade initiative also needs to foster public–private partnerships. As the private sector in developing countries (particularly LDCs) is mainly composed of SMEs, it faces a variety of constraints in enhancing its export capacity and competitiveness, ranging from inadequate finance and technology, inadequate infrastructure, high intermediary costs and regulatory red tape. Thus, enterprise development should form a key component of Aid for Trade. Further, the involvement of the private sector in aspects of the Aid for Trade Initiative, especially as regards production and exports, as well as infrastructure and institution-building, would be essential. This is particularly relevant as far as the development of trade facilitation infrastructure is concerned. The private sector can be a beneficiary of aid for trade, as well as a contributor in public–private partnerships. The private sector needs to be involved in aid for trade not only in delivery but also in the designing and planning stages.

It has been widely expressed that the current level of aid for trade is noteworthy but insufficient. There are important gaps in meeting the trade-related needs of developing countries. Thus, adhering to the principle of additionality and adequacy in aid for trade is critical, as highlighted previously.

There is also an important gap in terms of a coherent approach to aid for trade delivery. This applies to beneficiary countries, donors and international agencies. At the country level, coherence is needed for beneficiary countries to ensure mainstreaming trade into development objectives. This would require intra-governmental coordination, particularly between trade and finance institutions, as well as multi-stakeholder consultations, involving the private sector and civil society. This is a prerequisite for successful aid for trade, as needs and priorities, by virtue of country ownership, would have to be ascertained by beneficiary countries themselves. At the donor level, multilateral, regional and bilateral donors need to coordinate their efforts among each other and with beneficiary countries, as well as with implementing agencies, so as to map out developing countries' needs according to their national development strategies and to address well-defined areas in full coordination with each other and beneficiary countries. Such coherence is essential. It is critical that assistance be provided within the framework of the national development strategy and plan, not outside of it. At the international agency level, coherence and harmonisation of aid for trade deliverables remain crucial prerequisites for efficient results.

Like many other international development agencies, UNCTAD has experience in monitoring and evaluating technical assistance activities, based on a thorough needs assessment that identifies bottlenecks and crucial project-related variables. As part of its intergovernmental process, UNCTAD's technical assistance activities and specific projects are regularly monitored and evaluated by member States as well as by donors and beneficiaries. However, to engage in joint monitoring and evaluation of aid for trade projects, all parties involved (donors, beneficiaries, implementing agencies) must agree on the appropriate benchmarks that should be used to assess and monitor the performance of projects and their development impact.

Within the United Nations system, a growing number of programme-level evaluations rely on results-based frameworks. In the case of aid for trade, a results-oriented approach should be adopted from the outset, with evaluations conducted by external evaluators at various intervals of the Aid for Trade Initiative. Realistic targets and indicators of achievement should, to the extent possible, be identified with appropriate country-level management strategies and measurement tools to ensure country-level results.

Monitoring and evaluation will be required to ensure that aid for trade delivers the expected results. This could be implemented at two levels: (i) overall aid for trade disbursement and utilisation of resources and delivery of programmes; and (ii) specific aid for trade projects. A mechanism for annual reporting to a WTO body like the Committee on Trade and Development on overall implementation, in keeping with pledges made, might serve a useful role in this regard. OECD is assisting in monitoring such aid for trade flows. The first WTO Global Aid for Trade Review, drawing on regional reviews in different regions, was a positive exercise in monitoring and evaluation that deserves to be continued. In the context of its dialogue and consensus-building functions, UNCTAD could also provide input with regard to the overall monitoring and evaluation of aid for trade funding and implementation. It has contributed to the first WTO Global Aid for Trade Review and is a member of the WTO Advisory Board on Aid for Trade.

ANNEXES UNCTAD TECHNICAL COOPERATION

TABLE A.1.1. UNCTAD trust fund contributions, 2003–2006[a] (in thousands of dollars)

	2003	2004	2005	2006
Developed countries' contributions[b]	17 107	17 137	15 881	16 262
Developing and transition countries[c]	5 973	4 785	10 449	9 199
European Commission	2 173	691	2 888	2 343
United Nations system and other international organizations[d]	946	3 978	5 158	1 017
Private and public sectors	185	281	447	330
Total	**26 384**	**26 873**	**34 823**	**29 151**

[a] Excluding third-party cost-sharing contributions through UNDP.
[b] The figure for 2003 excludes $616,463 for associate experts; for 2004 it excludes $619,665 for associate experts; for 2005 it excludes $759,436 for associate experts; and for 2006 it excludes $809,055 for associate experts.
[c] The majority of these contributions are self-financing, for activities in the donor's own country, and are financed from proceeds of loans or grants from international financial institutions.
[d] For details see table 8 of the statistical annex (TD/B/WP/195/Add.2).

TABLE A.1.2. Total expenditure by UNCTAD on technical cooperation, and source of funds, 2003–2006 (in millions of dollars)

	2003	2004	2005	2006
UNDP	4.9	2.6	1.9	1.4
Trust funds	19.9	26.3	26.8	32.2
Regular budget and Development Account	3.0	1.7	1.7	1.7
TOTAL	**27.8**	**30.6**	**30.5**	**35.2**

TABLE A.1.3. Technical cooperation expenditures, by region and by programme, 2003–2006 (in thousands of dollars)

	2003	2004	2005	2006	
	Amount	Amount	Amount	Amount	%
Total	**27 788**	**30 594**	**30 485**	**35 223**	100.0
By region:					
Africa	3 842	4 929	5 103	6 985	19.8
Asia and the Pacific	5 503	5 056	5 975	6 841	19.4
Latin America and the Caribbean	2 239	2 138	2 265	2 887	8.2
Europe	906	1 360	1 155	1 241	3.5
Interregional	15 299	17 111	15 987	17 269	49.0
By programme:					
Globalization and Development Strategies	3 185	3 809	4 206	4 602	13.1
International Trade in Goods and Services, and Commodities	6 881	7 281	8 880	9 664	27.4
Investment, Technology and Enterprise Development	5 339	5 083	3 972	4 638	13.2
Services Infrastructure for Development and Trade Efficiency	9 066	10 783	11 182	14 150	40.2
Division for Africa, Least Developed Countries and Special Programmes	725	1 045	1 128	1 165	3.3
United Nations regular programme of technical cooperation (Section 22)	1 178	1 135	814	807	2.3
Direction of Management and Support Services	1 413	1 252	228	172	0.5
Technical Cooperation Service	–	206	75	26	0.1
Of which: least developed countries	*8 763*	*10 415*	*11 394*	*12 758*	*36.2*

TABLE A.1.4. Distribution of expenditures by interregional, regional and country activities, 2006 (In thousands of dollars and percentages)

	2006							
	UNDP[a]		Trust funds[b]		United Nations regular programme of technical cooperation and Development Account[c]		*Total*	
	Amount	%	Amount	%	Amount	%	Amount	%
Africa								
Country	199	0.6	4 858	13.8	–		5 057	14.4
Regional	97	0.3	1 831	5.2	–		1 928	5.5
Subtotal	***296***	***0.8***	***6 689***	***19.0***	–		***6 985***	***19.8***
Asia and Pacific								
Country	219	0.6	5 463	15.5	–		5 683	16.1
Regional	–	–	1 158	3.3	–		1 158	3.3
Subtotal	***219***	***0.6***	***6 621***	***18.8***	–		***6 841***	***19.4***
Latin America and the Caribbean								
Country	726	2.1	1 366	3.9	–		2 092	5.9
Regional	-	–	795	2.3	–		795	2.3
Subtotal	***726***	***2.1***	***2 161***	***6.1***	–		***2 887***	***8.2***
Europe								
Country	95	0.3	1 147	3.3	–		1 241	3.5
Regional	–	–	–	–	–		-	-
Subtotal	***95***	***0.3***	***1 147***	***3.3***	–		***1 241***	***3.5***
Interregional								
Interregional	29	0.1	15 578	44.2	1 662	4.7	17 269	49.0
Subtotal								
Interregional	29	0.1	15 578	44.2	1 662	4.7	17 269	49.0
Country	1 239	3.5	12 834	36.4			14 073	40.0
Regional	97	0.3	3 784	10.7			3 881	11.0
TOTAL	**1 365**	**3.9**	**32 196**	**91.4**	**1 662**	**4.7**	**35 223**	**100.0**

[a] UNDP-financed project.
[b] Voluntary contributions by member States and multilateral organizations.
[c] Sections 22 and 34 of the United Nations regular budget.

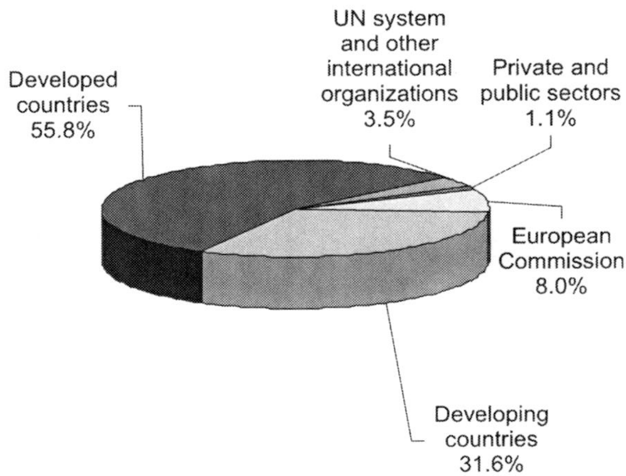

FIGURE A.1.1. Origin of trust fund contributions, 2006 (Per cent of total contribution)

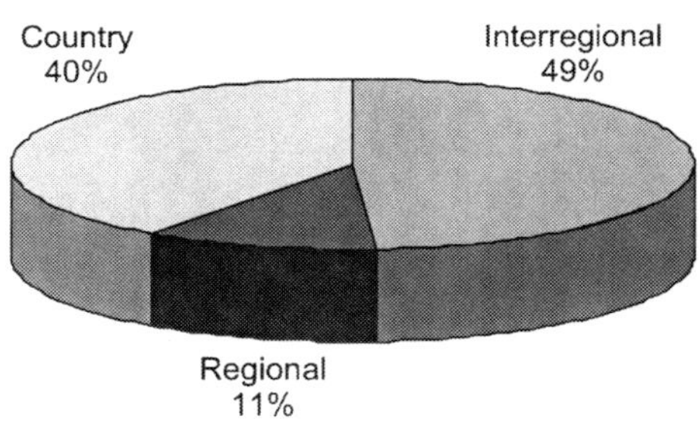

FIGURE A.1.2. Technical cooperation expenditures, by type of project, 2006 (percentage of total project expenditures)

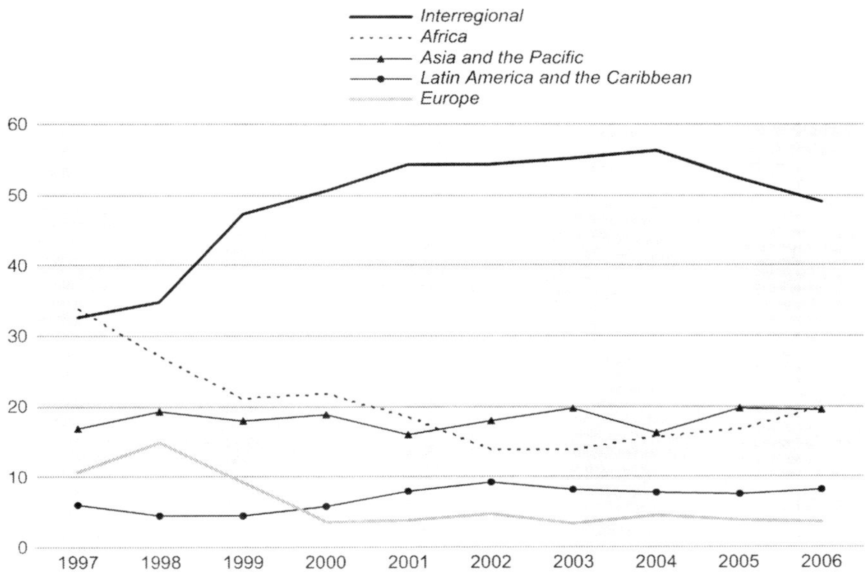

FIGURE A.1.3. Technical cooperation expenditures, by region, 1997–2006 (percentage of total yearly expenditure)

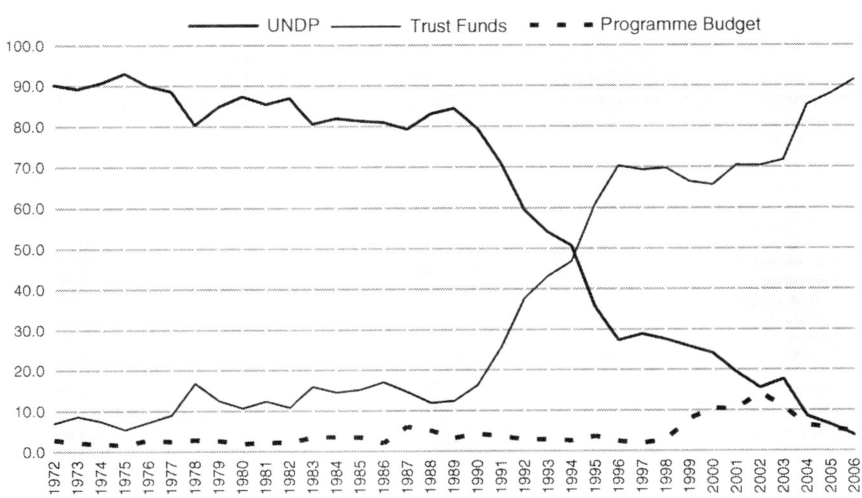

FIGURE A.1.4. Trends in mobilization of UNCTAD's technical cooperation, by source of funds, 1972–2006 (Per cent of total project expenditures)

TABLE A.1.5. Evolution of UNCTAD's technical cooperation, by source of funds, 2003–2006 (Actual project expenditures in thousands of dollars)

Year	UNDP[a] Amount	%	Trust funds[b] Amount	%	United Nations regular programme of technical cooperation and Development Account[c] Amount	%	**Total amount**
2003	4 887	*17.6*	19 919	*71.7*	2 982	*10.7*	**27 788**
2004	2 565	*8.5*	26 289	*85.2*	1 739	*6.3*	**30 594**
2005	1 943	*6.4*	26 816	*88.0*	1 726	*5.6*	**30 485**
2006	1 365	*3.9*	32 196	*91.4*	1 662	*4.7*	**35 223**

[a] UNDP-financed projects.
[b] Voluntary contributions by member States and multilateral organizations.
[c] Sections 22 and 34 of the United Nations regular budget.

TABLE A.1.6. Technical cooperation related to least developed countries, 2003–2006 (In thousands of dollars and percentages)

	2003	2004	2005	2006
(1) Country projects in LDCs	2 886	2 676	3 594	4 671
(2) Estimate of inter-country project expenditures directed towards LDCs[a]	5 877	7 739	7 800	8 087
(3) Total expenditure related to LDCs	8 763	10 415	11 394	12 758
(4) Total UNCTAD technical cooperation expenditure	27 788	30 594	30 485	35 223
(5) Share of LDCs in UNCTAD's technical cooperation, i.e. (3) / (4) *(percentage)*	32	34	37	36

[a] Secretariat estimate. This calculation was reached as follows: actual expenditure on inter-country projects concerning exclusively LDCs; plus for all other inter-country projects, 30% of interregional project expenditure; 60% of project expenditure in Africa; and 20% of project expenditure in Asia and the Pacific, and 5% of project expenditure in Latin America and the Caribbean, and the Arab States.

NOTES

1 Good summaries of the main arguments advanced in the aid literature can be found in Morrissey (2006).
2 A/RES/55/2.
3 A/RES/60/1.
4 Estimates for additional aid flows have been put forward in the context of meeting Millennium Development Goals, the Financing for Development Conference outcome, the recommendations of the Africa Commission, and pledges made at the Davos or G-8 meetings.
5 See for instance Burnside and Dollar (2004) on the critical role of institutions on the nature of aid-growth relationship.
6 For instance, Nilsson (1997) finds that \$1 of aid generates \$2.6 of exports from donor to recipient.
7 See Suwa-Eisenmann and Verdier (2007). These authors also identified a third channel through which aid flows may increase trade, namely by reinforcing bilateral economic and political links between recipient and donor countries.
8 The analysis of 'tied aid' is perhaps not very relevant for the current aid for trade process in the WTO context, as it is assumed that appropriate rules will be put in place to avoid such effects.

[9] Rajan and Subramanian (2005) found that aid has adverse effects on beneficiaries' competitiveness, primarily due to the real exchange rate overvaluation. The term 'Dutch disease' refers the theory that an increase in revenues from natural resources will, by raising the exchange rate, make the manufacturing sector less competitive, leading to the deindustrialisation of a country's economy.

[10] UN Millennium Project. March 2005. Investing in Development: A Practical Plan to Achieve the Millennium Development Goals. (New York).

[11] Ministerial Declaration of the Sixth WTO Ministerial Conference, 13-18 December 2005 (WT/MIN(05)/DEC, paragraph 57.

[12] See Statement by Supachai Panitchpakdi, Secretary-General of UNCTAD, to the WTO Aid for Trade Global Review, 20 November 2007 (www.unctad.org).

[13] WT/MIN(03)/12 (10 September 2003).

[14] See TD/429.

[15] TD/B(S-XXIII)/7(Vol. I).

REFERENCES

Burnside, A.C. and D. Dollar (2004). Aid Policies, and Growth: Revisiting the Evidence. World Bank Policy Research Working Paper, (3251).

Evenett, S. (2005). Some Tough Love On 'Aid For Trade', available online at http://www.evenett.com/articles.htm.

Francois, J. and M. Manchin (2007). Institutions, Infrastructure, and Trade, Development Studies, Working Paper No. 224, Centro Studi Luca d'Agliano.

Hoekman, B. and S. Prowse (2005). *Economic Policy Responses to Preference Erosion: From Trade as Aid to Aid for Trade*. Washington, DC: World Bank.

Hoekman, B. and M. Olarreaga (Eds.) (2007). *Global Trade and Poor Nations*. Washington: Brookings Institution Press.

Kemp, M.C. and S. Kojima (1985). The Welfare Economics of Foregn Aid. In G.R. Feiwel (Ed.), Issues in *Contemporary Microeconomics and Welfare* (pp. 470–483). London: Macmillan.

Laird, S. (2007). Aid for Trade: Cool Aid or Kool-Aid? G-24 Discussion Paper No. 48, Geneva: UNCTAD.

Morrissey, O. (2006). Aid or Trade, or Aid and Trade? Australian Economic Review 39, 78–88.

Nilsson, L. (1997). Aid and Donor Exports: The Case of the European Union. Lund Economic Studies, (70).

OECD (2006). The Development Dimension - Aid for Trade: Making It Effective, Paris.

Page, S. (2006). *Trade and Aid: Partners or Rivals in Development Policy?*. London: Cameron May.

Prebisch, R. (1964). Towards a New Trade Policy for Development, Report by the Secretary General of the United Nations Conference on Trade and Development, New York & Geneva.

Puri, L. (2005). Towards a New Trade "Marshall Plan" for Least Developed Countries: How to Deliver on the Doha Development Promise and Help Realize the UN Millennium Development Goals, UNCTAD Trade, Poverty and Cross-Cutting Development Issues Study Series No. 1, Geneva.

Rajan, R.G. and A. Subramanian (2005). Aid and Growth: What Does the Cross-Country Evidence Really Show?, IMF Working Paper No. 05/127, Washington: IMF.

Stiglitz, J. and A. Charlton (2006). Aid for Trade: A Report for the Commonwealth Secretariat, London.

Suwa-Eisenmann, A. and T. Verdier (2007). Aid and Trade. *Oxford Review of Economic Policy*, 23(3), 481–507.

UNCTAD (2006). Aid for Trade, United Nations: New York and Geneva.

Wilska, K. and M. von Bonsdorff (2006). Aid for Trade as a Vehicle for Enhancing Export Competitiveness, Paper prepared for the Seminar on Aid for Trade: Adjusting capacity to maximize sustainability in a liberalizing global economy, 14–15 November 2006, Glion, Switzerland.

UNCTAD (2007). Trade and Development Report 2007, Geneva: UNCTAD.

WTO (2006). Recommendations of the Task Force on Aid for Trade, WT/AFT/1, Geneva.

DAVID LUKE, LUCA MONGE-ROFFARELLO AND SABRINA VARMA, UNDP

CHAPTER 2

PERSPECTIVES ON AID FOR TRADE

2.1. INTRODUCTION

Aid for Trade (AfT) did not feature in the WTO Doha Ministerial Declaration in 2001 that launched the Doha Development Agenda, now often referred to as the Doha Round. It therefore falls outside the negotiating mandate that was agreed at Doha. However, as the negotiations progressed, it became clear that a major effort was required to provide assistance not only to build trade capacity to help poor countries take advantage of improved market access from a more developmentally-oriented Doha Round agreement but also to address supply side constraints and adjustment costs. To this extent it was understood from the start that AfT is a complement and not a substitute for new, fairer, trade rules.

Against this background, AfT emerged as a significant part of the WTO Hong Kong Ministerial Declaration in 2005 amid commitments from G7 and G8 Ministers to provide additional aid to developing countries "to build the physical human and institutional capacity to trade" and "to address adjustment related challenges and supply side constraints". The Declaration recommended the creation of a Task Force to examine how AFT could be operationalised and to consider how it could contribute to the development dimension of the Doha Round. The report of the Task Force was formally adopted by the WTO General Council in October 2006.

UNDP welcomed the AfT initiative as it was in line with its fundamental objective to promote human development as well as its mandate to support capacity development across a broad range of sectors, including trade. Indeed, UNDP's trade-related policy and advisory work placed special emphasis on the dynamic relationship between trade, growth and human development, and collateral measures and reforms with higher or positive human development outcomes in building supply side capacity.

But it was also recognised at UNDP that many of the elements that constituted AfT were not new. There were also concerns over the prospects for additionality and the lack of structure in the process for accessing the funds that were being pledged. The main AfT elements were in effect a re-packaging of instruments that have been part of both national development expenditures and external development assistance during the post-World War II period, while bringing renewed emphasis to trade-related development infrastructure and supply side constraints. On the other hand,

The usual disclaimer applies: the views expressed in this paper do not necessarily represent the official position of UNDP.

P. De Lombaerde, L. Puri (eds.), *Aid for Trade: Global and Regional Perspectives*,
United Nations University Series on Regionalism 2,
DOI 10.1007/978-1-4020-9455-2_3,

the inclusion in the initiative of 'adjustment' issues such as the fiscal, terms of trade, preference erosion and implementation costs associated with trade agreements had the effect of giving these issues more attention than they had previously received.

On the basis of this initial assessment, UNDP contributed its perspective on the questions under discussion to the AfT Task Force.(WTO, 2006a,b)[1] UNDP is a member of the Aid for Trade Advisory Body made up of IMF, ITC, OECD, UNCTAD, UNIDO, World Bank and the regional development banks. The Advisory Body was established by the WTO Director General in fulfilment of a key Task Force recommendation for a coordination forum consisting of major development partners to monitor and share experience on the operationalisation of the initiative.

UNDP is also a member of the OECD Technical Working Group on the Monitoring Framework on Aid for Trade, which includes key international organisations, regional development banks and selected developing country government representatives. And in its operations, UNDP maintains a substantive trade capacity development portfolio.

In the light of UNDP's association with the AfT initiative, this chapter has two main objectives. The first is to provide UNDP's perspective on the initiative and outline some concerns that may constrain the success of AfT as a vehicle for supporting a development-centred trade agenda in developing countries. And second, to identify the main elements of UNDP's support in the operationalisation of AfT.

2.2. UNDP'S ASSESSMENT OF THE AID FOR TRADE INITIATIVE AND EARLY CONCERNS

As previously noted, UNDP's initial assessment of the aid for trade initiative was that its main elements were not new. There were also concerns at UNDP over the ambiguity in regard to the additionality of resources as well as unclear procedures for developing countries to access these resources. UNDP's experience as a member of the OECD Technical Working Group on the Monitoring Framework on Aid for Trade also led to concerns over some aspects of the approach to monitoring that was developed during the spring of 2007. An elaboration follows.

2.2.1 Initial Assessment

During the post-independence years, various programmes and initiatives emerged at bilateral, regional and multilateral levels designed to enable developing countries, particularly low income ones, not just LDCs, to actively and effectively use trade as an instrument of development policy. These programmes include assistance in enhancing competitiveness including through support for enterprise development, diversification of exports, establishment of standards, and to deal with various human, institutional, infrastructure and other constraints (Malhotra, 2006).

Other types of programmes have also been implemented, although in a far less comprehensive manner. These have included support for adjustment difficulties (fiscal losses, terms of trade), as well as other costs incurred in the short, medium or long-term with respect to preference erosion (e.g. textiles and clothing, sugar, and bananas) and to implementation costs of trade agreements (e.g. trade facilitation, intellectual property, sanitary and phyto-sanitary standards, regulatory demands of services trade liberalisation).

Following from this, UNDP's assessment was that there are two broad sets of issues that aid for trade should help address but they are quite different in many important respects – in terms of objectives, essential characteristics, timeframes and financial sources for generating the resource requirements entailed.

The first group of aid for trade objectives are essentially developmental, necessary and desirable independent of trade negotiations and agreements, continuous in terms of need, primarily long-term in nature and clearly in the province of traditional "aid" as the source for the requisite financial resources. The resource implications have been and will be very substantial (e.g. the 2005 UK Commission for Africa report estimated that improving infrastructure in Africa could cost $20 billion). They should be met through an additionality of aid resources, as pledged in the 2005 G8 Gleneagles and UN Summits. They can be provided in the form of grants or as concessional loans, as has historically been the case. Overall, there is little which is conceptually new for this category of issues, even though a renewed emphasis on such issues can be useful and certain aspects (e.g. cross-border projects on infrastructure for landlocked countries, investments in projects addressing cross-country and regional impediments to trade development such as regional transport corridors, standards, disease or pest issues) have traditionally been neglected and need much higher priority than they have thus far been accorded. While all developing countries should be eligible for such support, prioritizing low income countries, as long as it is not exclusively restricted to them, can be viewed as justified.

It is the second set of aid for trade objectives and issues i.e. adequately and fairly addressing adjustment (fiscal loss of government revenue as a result of tariff reductions, changes in terms of trade for net food importers etc.), preference erosion and implementation costs of trade agreements that can be regarded as conceptually new, at least in terms of the discussion on how they should be treated. There is now broad agreement that there are 'winners' and 'losers' from trade negotiations. Developing country 'losers' are not just LDCs and low income countries but can also include middle income countries. The argument that aid for trade should be used to meet the various costs of trade agreements – adjustment, preference erosion and implementation to name the three most important – is relatively new in the mainstream discourse on trade or aid. There is also a strong case to treat these costs as different from the first group of AfT issues on almost all grounds: the objectives they are seeking to address, essential characteristics, the timelines involved, their one-off or transitory as distinct from continuous nature and the financial amounts, (considerable but lower than the first group of AfT), sources and terms on which they should be financed.

The case for using traditional forms of 'aid' or ODA budgets to meet adjustment costs is not strong. Since the costs for some countries are associated with gains for other countries, especially developed countries from trade liberalisation negotiated in trade agreements, aid criteria and existing aid budgets should not be used to compensate the 'losers' or pay for implementation costs of trade agreements. The need should also be met on a grant basis, not through concessional or other loans (because the costs borne by these countries result in benefit for other countries). Grants should also be provided to particular developing countries identified as 'losers', some of which may not be significant traditional beneficiaries of aid (e.g. Mauritius in view of its significant projected losses as a result of preference erosion).

2.2.2 Additionality

While the Task Force did not address details related to the funding aspects of AfT, recognizing the mandate given to the WTO Director General in the Hong Kong Ministerial Declaration to consult on 'appropriate mechanisms to secure additional financial resources', the report did underscore the importance of additionality and adequacy of funding to meets AfT needs.

A number of pledges were made in the lead up to, at and after the WTO Hong Kong Ministerial Conference in 2005, for increased aid towards trade related assistance by several donor countries, particularly the EU (2 billion euros per year by 2010 with half to be provided by the EC and the other half by EC Member States), US (USD2.7 billion per year for infrastructure and trade policy) and Japan (USD10 billion for over three years for trade, production and distribution related infrastructure). It remains unclear as to what extent the funds pledged are additional. The G8 commitment is for a total of $4 billion (including for the Enhanced Integrated Framework) annually[2]. However, analysis by commentators indicates hardly any additionality as the numbers show that there is 'little or no increase in total Aid for Trade and its share of total aid may fall' (Page, 2007). The overall amount of ODA is also projected to fall once the recent spate of debt write-offs comes to an end. On the other hand, the OECD argues that the money is not the issue and that there is enough money, but that the real challenge is to focus on aid effectiveness (OECD, 2006).

Additionality remains important for developing countries where there are concerns over resource diversion in the aid that they receive.

2.2.3 Accessing AfT

The Task Force recommended that AfT related projects and programmes should only be considered as such if they are identified as trade related development priorities in the recipient country's national development strategy. It fell short of recommending a new fund or multilateral financial arrangement for AfT. But the Task Force did acknowledge that there were existing gaps in ODA allocated to trade including lack of predictability in donor responses, inadequate support to infrastructure, adjustment, the productive sectors, and regional needs (WTO, 2006a,b).

Developing countries and other proponents of AfT on the other hand advocated for the establishment of a separate multilateral multi-year financial facility mainly to provide grants (supplemented by concessional loans limited to specific interventions), to ensure predictability and non-debt creating commitments. But there was no appetite among donor countries for establishing a dedicated AfT fund. Donors accordingly resisted this approach.

The compromise that was reached by the Task Force was that the established bilateral and multilateral procedures for accessing ODA resources will also be utilised for AfT. Efforts will however be made to make bilateral and multilateral funding more effective through improved mechanisms for coordination at country, regional and global levels based on the Paris Declaration on Aid Effectiveness. Incentives to enhance effectiveness were to be provided through a global monitoring process which will work to highlight and address the gaps on both the demand and supply sides, thereby building a more responsive relationship between demand from developing countries and response from the donor community.

But the specific procedures for accessing AfT resources by developing countries have yet to be determined. For example, it is not clear whether these countries will be required to follow some diagnostic method for assessing, establishing and costing their needs and priorities. It is also not clear which donor or group of donors will respond to which country and how to ensure equity between countries in the allocation of funds. Further, it is not clear whether priority will be given to specific groups of countries such as LDCs and small vulnerable economies (SVEs) – two categories of developing countries recognised at the WTO – or whether allocation will be on first come first served basis.

2.2.4 Aid for Trade Monitoring and Evaluation Framework

The 'big idea' of the Task Force was its recommendation for the establishment of a global AfT monitoring and evaluation (M & E) framework. This was primarily to encourage and build confidence in the AfT initiative as well as provide an incentive to developing and donor countries and the relevant international organisations to work together to ensure that AfT responds to the issues and challenges on the ground, with concrete and visible results. In the light of the studied ambiguity over the additionality of resources and the lack of clarity in the compromise adopted by the Task Force that existing bilateral and multilateral channels will be utilised for accessing AfT funding, it is not surprising that M & E became the centre-piece of its recommendation.

This required the Task Force to elaborate precisely what is to be monitored and evaluated. It defined AfT activities as those which have been identified as trade-related priorities in the recipient country's national development strategies, such as the PRSP. Building upon the definitions used in the Joint WTO/OECD Database, it recognised that these activities fell into six major categories, namely (a) trade policy and regulations (b) trade development (c) trade-related infrastructure (d) building productive capacity (e) trade-related adjustment and (f) other trade-related needs.

The Task Force envisaged that M & E will not only focus on issues of effectiveness and efficiency at country and regional levels including progress and results in project and programme implementation but also donor commitments, agency support and strategies for meeting them.

Tip-toeing carefully around the acknowledged fact that the WTO is not a development agency, the Task Force however recommended that that the WTO should become the 'clearing house' for AfT M & E information and a forum for a global periodic review and assessment of what was being achieved under the initiative. In addition to the global review, the Task Force recommended expanding the role of the WTO Trade Policy Review Mechanism – which assesses the trade policies of member states on a periodic basis – to include assessment of the effectiveness of AfT support provided by donor member states and progress in developing trade capacity in AfT recipient member states.

Following the adoption of the Task Force Report in October 2006, the OECD was invited to take the lead in elaborating the M & E framework with the help of a Technical Working Group (TWG) on the Monitoring Framework on Aid for Trade. The TWG consisted of representatives of the main development agencies, regional development banks and selected developing countries. Guided by the recommendations of the Task Force, the TWG agreed on three main strategies for generating M & E information: (i) measuring global aid for trade flows using the OECD-DAC creditor reporting system (CRS), the main source for the Joint WTO-OECD Database which had been available since 2001; (ii) measuring global aid flows at the donor and agency level based on self assessments of current activities and fulfillment of pledges; and (iii) reporting by recipient countries of their specific AfT experience.

UNDP identified five main shortcomings with the approach adopted by the TWG and articulated them at the various meetings and consultations of the group that were held during the spring of 2007.

First, while the Joint WTO-OECD Database represents an indispensable source of information on AfT flows, it remains essentially an inventory of trade-related technical assistance and capacity building initiatives and falls far short of providing an assessment of AfT quality, effectiveness and on-the-ground impact and results.

Second, the limitation of the Database is compounded by the fact that there are no common benchmarks, criteria, or indicators to determine the adequacy or additionality of AfT flows. Indeed, the Task force had acknowledged that "clear and agreed benchmarks are necessary for reliable global monitoring . . . to ensure accurate accounting and to assess additionality" (WTO, 2006a,b). No global benchmark has been established. On the contrary, the M & E framework adopted by the TWG only requires donors and agencies to undertake self-assessments. As each donor organisation will provide its own year-on-year financial and other reporting, this is unlikely to shed much light on additionality let alone on impact and results. The reporting required is more likely to be process oriented.

Third, the TWG missed an opportunity to provide for independent assessments of the impact of AfT flows through in-depth country case studies of how effectively they contributed to a recipient country's trade and development objectives and towards integration into the global trading system. This also requires assessing the performance and effectiveness of donors and the capacity of recipient countries, together with identifying what categories of AfT and modes of financing – concessional, non-concessional, and other combinations – were most effective. As experience is gained with the M & E framework, it is clear that better tracking of performance will be required to provide the necessary incentives to the partners involved in the initiative to produce results.

Fourth, the TWG also missed an opportunity to provide for the regular analysis of specific thematic issues to enhance understanding of the complexity of trade capacity building on the ground. Examples of such issues are the institutional capacity of national trade ministries or impact of pro-poor trade policies.

Fifth, despite the inclusion of regional initiatives in AfT, the TWG did not adequately clarify how regional AfT will be measured and evaluated. It is also not clear how private sector support is to be monitored.

2.3. UNDP'S SUPPORT PORTFOLIO

Notwithstanding UNDP's concerns, the AfT initiative has been welcomed and embraced as its fundamental objectives are in line with UNDP's corporate concerns to promote human development and support capacity development across the main economic and social sectors, including trade. It is also recognised that AfT is an evolving initiative that seeks to strengthen both the demand for AfT from developing countries and the donor response within the framework of the Paris Declaration. It is anticipated that the initiative would be flexible enough to correct its shortcomings as experience is gained.

Fostering inclusive globalisation has been adopted as a key results area in UNDP's Strategic Plan 2008-11. UNDP's strategy to support the AfT initiative including the Integrated Framework for Trade-related Technical Assistance (or IF) is guided by its corporate mandate and has the following specific objectives:

- Supporting trade mainstreaming in national development/poverty reduction plans
- Building capacity to negotiate, interpret and implement trade agreements
- Enhancing competitiveness and overcoming supply side constraints. While the main focus of UNDP's support is at the country level with national priorities and national ownership providing the context and framework, addressing cross-border issues is emerging as an important part of UNDP's regional trade programmes in Africa, Asia and the Pacific, Arab states and the countries in transition.

UNDP's trade capacity development support for the most part falls under the following AfT categories: (a) trade policy and regulations (b) trade development and building productive sector capacity.

2.3.1 Trade Policy and Regulations

Specific support activities carried out by UNDP under this category include trade diagnostics and research, policy analysis, trade mainstreaming in national development/poverty reduction strategies, training, and institutional development. Support through global, regional and country programmes is estimated at between US$8-10 million a year for approximately 60 interventions in over 40 countries.

UNDP's regional bureaus and country offices collaborate with its Bureau for Development Policy which maintains a Trade and Human Development Unit in Geneva under the Inclusive Globalisation Cluster of the Poverty Group to implement various trade policy capacity development interventions with country ownership as a guiding principle. Approximately 40 staff are actively engaged either full or part-time in supporting trade capacity issues with the support of UNDP's poverty reduction practice community which includes over 300 economic development/poverty reduction specialists and maintains an e-knowledge network to facilitate information exchange and capacity building.

As recognised by the Task Force, strengthening the 'demand side' of AfT requires donor and agency support for country ownership and country-driven approaches on the one hand, and, on the other, a commitment of governments to fully mainstream trade into their national development strategies. In some countries, the process for mainstreaming trade, formulating trade strategies, and proposing priority trade projects and programmes for donor financing needs to be strengthened through technical assistance and capacity building to help put in place effective and sustainable trade policy frameworks and processes. UNDP works alongside country stakeholders, donors and international agencies to provide and coordinate support for strengthening the 'demand side'. In particular, through the United Nations Resident Coordinator System, UNDP helps to facilitate greater donor and agency coordinatation and harmonisation of procedures to ensure coherence at both the operational and policy levels. In addition, at country and global levels, UNDP has provided trust fund management services to facilitate pooled funding approaches. UNDP serves as the global trust fund manager for the IF.

2.3.2 Trade Development and Building Productive Sector Capacity

Support for trade development and building productive sector capacity is provided mainly under UNDP's private sector portfolio which seeks to foster inclusive markets and consists of two broad type of interventions: private sector development aimed at increasing the contribution of micro-, small and medium sized enterprises to economic growth and poverty reduction; and private sector engagement aimed at fostering partnerships with a range of companies from multinationals (northern and southern), to small and medium sized enterprises (SMEs), often with a broader range of development objectives in mind.

The private sector development portfolio accounts for approximately US$80 million of programme spending annually on over 400 projects in 101 country offices. It comprises a wide range of predominantly 'single theme' private sector development interventions, including support for business development services, access to microcredit, and entrepreneurship development. Single theme projects of this kind account for approximately 57 per cent of the portfolio in terms of simple project headcount. A further 25 per cent goes to sectoral and supply chain projects of various kinds, and approximately 11 per cent to policy related interventions, mainly in the area of improving the business environment.

Private sector engagement initiatives generate approximately USD20 million per year in private sector funding contributions to UNDP through approximately 130 programmes in 75 country offices and regional centres. Cost sharing private sector partnerships are a relatively recent phenomenon in the development community at large but performance to date suggests that UNDP is extremely well positioned in this area with a strong and rapidly growing portfolio. Approximately 60 per cent of partnership projects are in the area of poverty reduction, followed by energy and environment and the social sectors. The majority is motivated by Corporate Social Responsibility (CSR) objectives and consequently includes a wide spectrum of interventions ranging from philanthropic donation s to market-led investments.

UNDP's capacity development and democratic governance practice communities are also active on private sector support which is mainly carried out through the Bureau of Partnerships and the UNDP affiliate, the UN Capital Development Fund (UNCDF), in collaboration with regional bureaus and country offices.

2.4. CONCLUSION

While globalisation has been a key driver of economic growth benefiting hundreds of millions of people, these benefits have not been shared equitably either between or within countries. Well formulated domestic policies will not be as effective unless the globalisation process is beneficial to all developing countries. To this extent, AfT could bring significant benefits to developing countries in providing assistance not only to meet adjustment costs but also to take advantage of market access opportunities from the international trading system. This can impact positively on sustainable economic growth and increasing levels of human development in the drive to achieve the MDGs.

It was recognised at UNDP that many of the elements that constituted AfT were not new. There were also concerns over the prospects for additionality of AfT resources, unstructured access to these resources, and gaps in the M & E framework that has been put in place. Yet this is also an evolving initiative that seeks to strengthen both the demand for AfT from developing countries and the donor response within the framework of the Paris Declaration. As a multilateral partnership, there are good grounds for optimism that the flaws would be addressed and resolved over time as experience of the initiative is gained. For its part, UNDP – as a member of the WTO Advisory Body and OECD Technical Working Group on the Monitoring Framework, and with its own substantive trade capacity support portfolio – is committed to ensuring the successful operationalisation of AfT.

NOTES

[1] UNDP also participated in a number of Aid for Trade related conferences and meetings during 2006–2007 organised by the OECD, UNCTAD, the Commonwealth Secretariat, AITC and the WTO. UNDP is an observer on the WTO Committee on Trade and Development (CTD) where aid for trade is a regular feature on the agenda and participates in the governing bodies of the Integrated Framework, a programme which has overlapping objectives with the aid for trade initiative.

[2] See G7 Finance Ministers, "Conclusions on Development", London, 4–5 February 2005; G8 Summit Gleneagles Communiqué, "Africa, Promoting Growth", Gleneagles, Scotland, 6–8 July, 2005.

REFERENCES

Malhotra, K. (2006). Aid for Trade Concept Note, UNDP, New York, February 2006.

OECD (2006). The Development Dimension, Aid for Trade: Making it Effective, Paris.

Page, S. (2007). The Potential Impact of the Aid for Trade Initiative, G-24 Discussion Paper Series, No. 45, Geneva, UNCTAD, April 2007, p. 27.

UK Commission for Africa (2005). Our Common Interest: Report of the Commission for Africa, March 2005, London.

WTO (2006a). Recommendations of the Task Force on Aid for Trade, WT/AFT/1, 27 July 2006, Geneva.

WTO (2006b). Summary of Contributions from Inter-governmental Organisations: Aid for Trade Task Force, WT/AFT/W/17, Geneva.

PHILIPPE DE LOMBAERDE, UNU-CRIS, AND GEORGE MAVROTAS, GDN

CHAPTER 3

AID FOR TRADE, AID EFFECTIVENESS AND REGIONAL ABSORPTION CAPACITY

3.1. INTRODUCTION

The aim of this chapter is to place the Aid for Trade (AfT) discussion in the broader context of the current debate on international aid flows. At the same time, we want to highlight the growing importance of the regional level in the whole discussion on aid, as clearly illustrated by the case of AfT where aid flows are explicitly linked to region-to-region trade negotiations, requiring regional donor coordination, on the one hand, and regional absorption capacity, on the other.

From a broader perspective on foreign aid, recent years have witnessed a changing landscape and new issues have emerged partly due to the ongoing debate on how the Millennium Development Goals can be achieved by 2015 using existing sources of finance (predominantly foreign aid) but also other sources of development finance. At the same time, regionalization revealed new challenges for development aid. Against this background the chapter will attempt to discuss the emerging new scene for foreign aid and development finance in general, and the key issues involved in the area of development cooperation and regionalization that are relevant to contextualize the Aid for Trade discussion.

3.2. FOREIGN AID AND DEVELOPMENT FINANCE: RECENT DEVELOPMENTS

Recent work in the broad area of development finance but also in the area of international aid flows has stressed the growing complexity of the multilateral development finance system and more importantly that the current multilateral development finance 'system' is rather a *non-system*, in the sense that unlike some of its elements (such as the Bretton Woods sister organisations) it has not been planned, and it is not the result of coherent design (Mavrotas and Reisen, 2007). It can be coined a non-system, very much like Max Corden did for the international monetary non-system.

Donor fragmentation has also become a central issue in the area of international aid (Knack and Rahman, 2004:2207) recently.[1] A recent World Bank study has underlined that "the average number of donors per country nearly tripled over the last half

The views expressed in the chapter are those of the authors alone and should not be attributed to the United Nations, United Nations University or the Global Development Network.

P. De Lombaerde, L. Puri (eds.), *Aid for Trade: Global and Regional Perspectives*,
United Nations University Series on Regionalism 2,
DOI 10.1007/978-1-4020-9455-2_4, © Springer Science+Business Media B.V. 2009

century, rising from about 12 in the 1960s to about 33 in the 2001–2005 period. The combination of more bilateral donors and of an increasing number of multilateral channels has led to an increasingly crowded aid scene. Aid channel proliferation at the country level has been substantial, particularly after the end of the Cold War when the number of countries with over 40 active donors and international organizations grew from zero to thirty-one. The number of international organizations, funds and programs is now higher than the number of developing countries they were created to assist" (World Bank, 2007:19). Figure 3.1 clearly illustrates the complexity of aid channels from the recipient country perspective.[2]

The growing complexity of the development finance system is also captured well by Figure 3.2.

At the same time, recent years have witnessed key landmark events in the multilateral development finance which include the adoption of the Millennium Development Goals (MDGs) in 2000, the Monterrey Consensus in 2002, the endorsement of the Paris Declaration on aid effectiveness and managing for development results in 2005, and the G8 commitments at Gleneagles in 2005 to increase Official Development Assistance (ODA) to Africa by $25 billion a year by 2010 (reiterated in the G8 Meeting at St Petersburg in 2006). Finally, 2008 is also expected to be a very special year for development aid and development finance in general, in view of the *Third High-Level Meeting on Aid Effectiveness* in September 2008, where heads of governments,

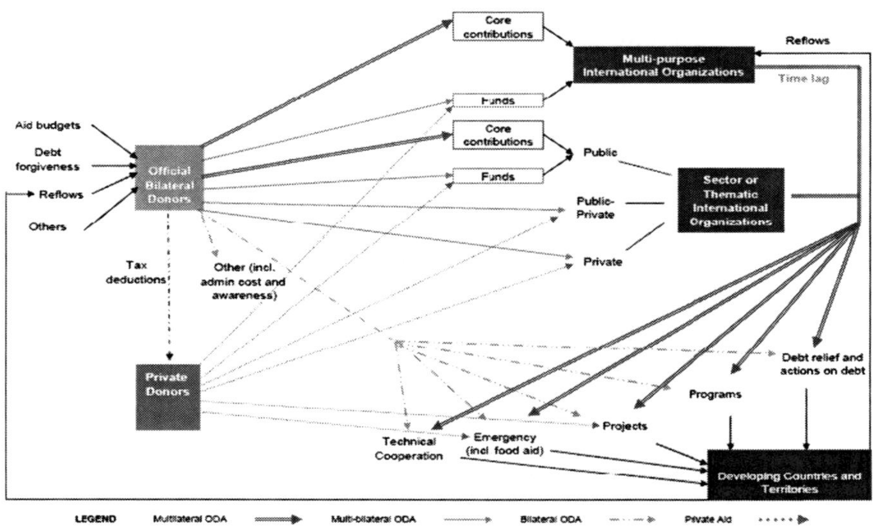

FIGURE 3.1. The complexity of aid channels from the recipient country perspective (*Source*: World Bank (2007), Aid Architecture: An Overview of the Main Trends in Official Development Assistance Flows)

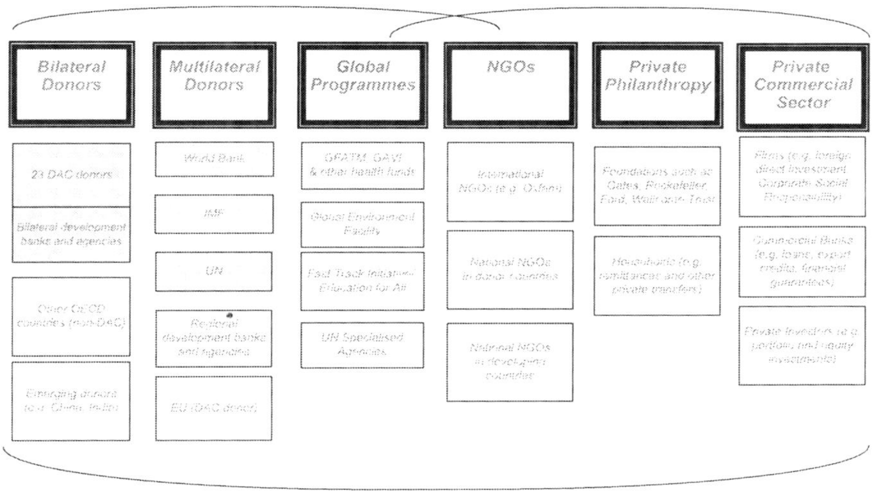

FIGURE 3.2. The complexity of development finance system (*Source*: OECD Global Forum on Development (2006) and Mavrotas and Reisen (2007))

multilateral institutions and civil society organisations gathered in Accra to review the progress made in implementing the Paris Declaration. The Accra Forum will be closely followed by the *Follow-up International Conference on Financing for Development* to review the implementation of the Monterrey Consensus in Doha at the end of the year.[3,4]

There is now widespread agreement that more aid is required to meet the MDGs but at the same time improving dramatically aid effectiveness, mobilizing domestic resources and exploring new sources of development finance beyond aid, are all crucial factors. In this context, the international community has tried also recently to explore other possible sources of funding beyond aid in view of the fact that substantial time may be needed for donors to double aid (and recent delays in aid increases clearly support this line of argument).[5]

Finally, regionalization revealed *inter alia* new challenges for development aid in particular and development finance in general. On the one hand, regionalization can offer new opportunities for regional financial cooperation in the area of development assistance as has recently been reflected in new regional initiatives such as the New Partnership for Africa's Development (NEPAD). On the other hand, regional cooperation in the area of development finance can have profound implications for the existing foreign aid architecture. Against this background the chapter will attempt to discuss the emerging new scene for foreign assistance, the key issues involved in the area of development cooperation and to reflect on recent regional initiatives such as NEPAD.

3.3. CENTRAL ISSUES IN THE CONTEXT OF THE NEW LANDSCAPE FOR DEVELOPMENT AID

In this rapidly changing environment for development finance it is not surprising that issues of scaling-up aid and aid effectiveness have dominated the policy dialogue in recent years and the relevant literature has blossomed.

3.3.1 Aid-Growth Empirics: Where Do We Stand?

The aid and growth empirics are as old as aid giving itself. Indeed, for almost five decades numerous empirical studies have tried to address in various ways the impact of aid on growth in aid-recipient countries and thus shed light on the crucial aid effectiveness issue. For many years the empirical literature on the effectiveness of aid remained inconclusive partly due to lack of good data on aid but also partly due to inappropriate econometrics and simplistic empirical specifications employed in most of the empirical studies. Recent years, however have witnessed important changes in the aid arena[6] which *inter alia* revived the interest in aid effectiveness issues.

A plethora of empirical studies which were published over the last decade, employing better aid data and more robust econometric techniques, have clearly improved our knowledge regarding the impact of aid on growth and poverty reduction. Although the literature on the empirics of aid still remains far from conclusive, a substantial part of the new generation of aid studies seem to suggest that overall aid seems to have a positive impact on growth.

It is clearly beyond the scope of the present chapter to review the vast literature of the aid-growth empirics. See Tarp (2000), Beynon (2002 and 2003), Hudson (2004), Dalgaard *et al.* (2004), Collier and Dollar (2004), Addison *et al.* (2005a,b), Doucouliagos and Paldam (2005), Radelet (2006), McGillivray *et al.* (2006), and very recently, Easterly (2006, 2007), Bourguignon and Sundberg (2007a,b), Riddell (2007) and Lahiri (2007) on recent assessments of the aid effectiveness literature; see also Mosley (1987), White (1992) and Cassen (1994) on earlier reviews of the literature.

It has to be stressed that the turning point in the aid-growth empirics was the Burnside and Dollar (2000) seminal paper published in the *American Economic Review* in 2000 (and in fact its earlier version published as a World Bank Working Paper in 1997 by the same authors as a background paper for the World Bank study *Assessing Aid* (1998)). One of the key conclusions of the Burnside and Dollar (2000) paper was that aid works better in countries with sound policy regimes and more precisely that "....aid has a positive impact on growth in developing countries with good fiscal, monetary, and trade policies but has little effect in the presence of poor policies" (Burnside and Dollar, 2000:847). It is notable that the above study has been very influential among donor agencies since it provided the donor community with an important policy criterion for allocating aid, namely that aid should be allocated on a selective basis to those countries that have adopted good policies in view of the central finding of the study that aid works only in a good policy environment.[7]

At the same time, the Burnside and Dollar study has mobilized a relatively large and still growing empirical literature in recent years trying to delve deeper in the aid-policies-growth nexus emphasized in the Burnside and Dollar paper. Most of these studies have seriously questioned the validity of the empirical results (and thus the policy lessons emerging from the Burnside and Dollar study) on many grounds such as inappropriate econometrics, problematic definition of the 'policy' variable, inappropriate specification of the empirical model, endogeneity issues etc.[8]

In a recent critique of the Burnside and Dollar paper, Easterly et al. (2004), by adding 4 more years (1994–1997) to the original Burnside and Dollar dataset convincingly show that the Burnside and Dollar finding regarding the aid-policy interaction is not robust to the use of this additional data thus casting serious doubts on the policy implications emanating from the Burnside and Dollar study. Antipin and Mavrotas (2006) take a fresh look at the aid-growth empirics by adding a new methodological dimension to the Burnside and Dollar paper and the overall aid-growth empirics. They use three different data sets (including the one used in the Burnside and Dollar paper) and Bayesian instrumental variable methods to test the robustness of the central finding of the Burnside and Dollar paper related to the aid and policy interaction coefficient. In doing so, they apply Bayesian instrumental variable techniques to find the most probable parameter values in the growth equation. A central finding was that the problematic interaction term of aid and policy is not statistically significant in the model even with the heteroscedastic-consistent estimator, and most importantly, its marginal effect on real per capita GDP growth is substantially smaller than in the Burnside and Dollar (2000) paper. This obviously raises important questions regarding the robustness of the Burnside and Dollar findings and the validity of the crucial policy implications emerging from the above study. Finally, Baliamoune-Lutz and Mavrotas (2008), add a new dimension to the aid-growth empirics by examining the impact of institutional quality and social capital on aid effectiveness. They find strong evidence that social capital and institutions enhance aid effectiveness. Moreover, once they account for the role of social capital and institutions, the impact of policies tend to disappear. Needless to say, these findings have important policy implications as they indicate that conditioning aid allocation on 'good policies' may not lead to an optimal (or fair) allocation of aid, as countries with high social capital at the macro level could actually make good use of aid regardless of the quality of policy. And, this also casts doubt on the conclusions in Burnside and Dollar (2000) and the policy lessons derived from their findings.

3.3.2 Increasing Aid: Implications for Macroeconomic Management

As already mentioned, recent calls for increasing aid generated a new interest among researchers and policy makers regarding the macroeconomic implications of such a 'big-push' approach. At the same time, there exists an ongoing debate whether additional aid could be absorbed effectively, while issues related to aid volatility, aid heterogeneity, fiscal response to increased aid inflows and those related to political economy aspects of aid[9], have been under scrutiny recently.

A central issue seems to be the capacity of the aid-recipients to absorb such large amounts of development assistance and whether the right institutional mechanisms are in place to facilitate the absorption of additional aid.[10] A message emanating from recent work is that it is not sufficient to scale up aid efforts by raising and transferring more money (see Killick and Foster, 2007; Bourguignon and Sundberg, 2007a; Mavrotas and Nunnenkamp, 2007). At least part of the blame for not making a lot of progress with the MDGs so far falls on insufficient targeting of sector-specific aid; in this context, unless aid is better targeted, scaling up aid is unlikely to have the desired effects (Thiele et al., 2007). And this issue is also related to the neglect of the heterogeneous character of aid in the vast literature of the aid empirics. Furthermore, the 'big push' and 'absorptive capacity' approaches cannot be reconciled without a reform on the aid architecture associated with recent calls for scaling up aid (Guillaumont and Guillaumont-Jeanneney, 2007). This also echoes Maxwell (2002) who suggests that it would be a missed opportunity to increase aid without considering the apparatus for delivering such large amounts of money. Finally, we should bear in mind that the ultimate measure of aid effectiveness is how aid affects the lives of poor people in developing countries; and the voluminous literature on aid's macroeconomic impact has remarkably little to say on this topic, and less still in terms of practical advice to government officials and aid administrators on how to improve the effectiveness of development aid. In this context focussing on impact evaluation studies might be the right avenue for practical and policy-focussed research in this area (White, 2008).

A related issue concerns the impact of aid on the fiscal sector in the aid-recipient economy. This issue attracted a lot of attention by researchers and policymakers alike in the 1990s with the emergence of a new theoretical and empirical literature known as the 'fiscal response literature'[11] building on the pioneering paper by Heller (1975). A separate strand of this literature deals with the fungibility issue, namely that the aid-recipient government may reduce its own resources from the sector which receives aid and channel them to other sectors of the budget.[12] Needless to say, fiscal response issues are becoming quite timely now in view of the need to scale up aid to accelerate progress with the MDGs.

Another issue which attracted much of attention recently in academic and policy circles is related to the management of aid volatility, and in particular the unpredictability of aid flows, of crucial importance for the attainment of the MDGs.[13] In this regard, Pallage and Robe (2001) report evidence according to which aid flows are highly volatile over time (on average, two to three times as volatile as the recipient's output) and overwhelmingly procyclical. Bulíř and Hamann (2003), by using data covering 72 countries over the period 1975–1997 focus on a comparative analysis of the volatility of aid flows and the volatility of domestic revenue in aid recipients, rather than on the volatility of aid inflows *per se*. They found that aid flows are more volatile than domestic fiscal revenues and tend also to be pro-cyclical, and that there are much larger prediction errors in programme assistance than in project aid, and a stronger tendency to over-estimation. Fielding and Mavrotas (2005), examined aid volatility using data for 66 aid recipients over the period 1973–2002. They tried to improve upon earlier work in this important area by disaggregating total aid inflows into sector and programme aid. In this way they avoid focussing on a single aggregate, unlike previous studies on aid volatility. They found that *inter alia* the institutional quality of the aid recipient affects the stability of sector aid but not that of programme assistance, and that macroeconomic stability affects the stability of both kinds of aid, as does the extent to which a country relies on a small number of individual donors. More recently, Hudson and Mosley (2006), by revisiting the Bulir and Hamann's analysis of aid volatility argue that there are problems with both the data and the methodology employed in the above study. Similarly to Fielding and Mavrotas (2005) they disaggregate aid by aid donor and type and construct an empirical model which seeks to explore the causes and effects of aid volatility. Fielding and Mavrotas (2008) examine aid volatility by constructing a panel of over 60 aid recipients for the period 1975–2004. Their results are generally robust to differences in the way aid measures are deflated, and to differences in the way the data are detrended. They found that the variability of aid growth in a particular country depends on the characteristics of both the donors and the recipients. Macroeconomic stability in the recipient country, in terms of low inflation, reduces total aid volatility, but has no significant impact on sector aid volatility. In a similar way, countries that have recently agreed to IMF conditionality experience higher total aid volatility, but not higher sector aid volatility. The volatility of both total and sector aid also depends on the quality of political institutions, although the dimensions of institutional quality that impact on sector aid volatility appear to be different from those impacting on total aid volatility, for reasons that are as yet unclear. Finally, Chauvet and Guillaumont (2008), argue (by somewhat diverging from the current concern about aid volatility) that aid has a stabilizing impact first with respect to exports volatility, and second and more generally as a dampening factor of income volatility. The authors show that aid volatility may lower and possibly cancel this effect when it is pro-cyclical with regard to exogeneous shocks, and even, but more hardly, when it is contra-cyclical and very high compared to other sources of shocks. Even if it is pro-cyclical aid is stabilizing with respect to exports, if its volatility is lower than that of exports. And when it is contra-cyclical, it is stabilizing only as far as its volatility does not exceed a certain threshold.

Finally, issues associated with the heterogeneity of aid flows and the need to look at disaggregated aid figures rather than on aggregate ones in empirical work have gained momentum in recent years with the emergence of a series of papers in this area. The central argument here is that the use of a single figure for aid, a typical feature of the aid effectiveness literature, cannot capture this aid heterogeneity, thus leading to aggregation bias in the empirical 'evidence' reported (see Cassen, 1994; White, 1998, Mavrotas, 2002a,b; Clemens *et al.*, 2004; Mavrotas, 2005; Ouattara, 2006; Mavrotas and Ouattara, 2006a,b). For more recent contributions to this also policy influential area see the Special Issue edited by Mavrotas and Nunnenkamp in the *Review of World Economics (Weltwirtschaftliches Archiv)* in 2007 and the papers therein.

3.4. NEW REGIONALISM

In the ongoing process of adjustment and reconfiguration of the global governance architecture, the strengthening of the (supra-national) regional level is probably one of the more important factors. This phenomenon is often referred to as the 'new regionalism' or the new wave of regionalization.

This phenomenon is often understood as a primarily, or even exclusively, quantitative phenomenon. And to illustrate this, usually reference is made to the growth in the number of regional trade arrangements (RTAs) notified to the GATT/WTO. Since about the end of the 1980s we have indeed witnessed a marked surge in the number of notifications per year. More than 250 RTAs have already been notified and it is estimated that more than 50 other arrangements are operational although not (yet) notified.[14] This new wave of regional arrangements is usually explained by the uncertainty about the outcome of the Uruguay Round at that time but it is now clear that the rise in regionalism is a longer term phenomenon linked to more structural factors related to scale and mobility in the context of globalisation.

However, this quantitative expansion covers a diverse reality and the new regionalism should not be reduced to it. One should distinguish between at least four meanings of new regionalism. In addition to the quantitative definition (the growing number of regional integration agreements - RIAs), the new regionalism refers also to qualitatively different RIAs that were launched in that same period (covering also non-trade issues, forms of border-zone integration, etc.). In a third definition, the new regionalism refers to new theories of regional integration (explaining, for example, the link between regionalism and globalisation), and finally, the new regionalism also refers to new methodological approaches (rediscovering comparative studies, emphasising the need for interdisciplinary approaches, etc.) (Hettne et al., 1999–2001; Breslin et al., 2002; De Lombaerde, 2003; Gavin and De Lombaerde, 2005; Farrell et al., 2005).

Although the regionalist move can be observed on a global scale, it should further be acknowledged that there is no unique model of regionalisation (Fawcett and Hurrell, 1995). A variety of governance models exist, responding to different sets of external forces and, especially, different intra-regional factors and circumstances.[15] The well-known Balassa typology of regional economic integration does not capture the actual variety of forms of regional cooperative behaviour. Hettne and Söderbaum (2004), for example, present a typology of regional cooperation mechanisms based on whether cooperation is delivered by an organization or network, on the one hand, and on whether cooperation is unidimensional or multidimensional. Regional development banks, security organizations, development corridors, research networks, river basin organisations or (just) common policies are all examples of institutionalised regional governance. According to the authors, there is a tendency observable from mainly unilateral forms of cooperation (often at the level of organizations) towards multidimensional and hybrid forms of cooperation.[16] These tendencies imply increased numbers of actors involved in the process and more heterogeneity among them. From a longer term perspective, this process of regionalization can be described in terms of levels of 'regionness' — i.e. successive orders of regional social space, system, society, community and institutionalized polity. Increasing regionness implies that a geographical area is transformed from a passive object to an active subject — an actor — increasingly capable of articulating the transnational interests of the emerging region (Hettne and Söderbaum, 2000).

In the economic sphere, new regionalism also means a deepening of economic integration, expanding the process to cover non-trade (i.e. regulatory) issues (compatible with, exceeding or conflicting with WTO regimes and agendas)[17] and –in some cases- monetary integration. Experiences with monetary integration (regionally fixed exchange rates, common currencies, dollarisation) show mixed results but are likely to become more important in the future (Cuyvers et al., 2005). Recent initiatives include, for example, the expansion of WAEMU in West Africa through a plan for monetary integration involving both francophone and anglophone states in West Africa, set out by the ECOWAS heads of state in 1999; the inclusion of a future monetary union in the treaty creating the East African Community (EAC) Customs Union, ratified in 2004; the dollarisation in Ecuador (2000) and El Salvador (2001); and the inclusion of an ASEAN common currency in the 2020 roadmap, adopted in 2003 at the Manilla Summit. Next to the deepening of economic integration schemes, in a number of cases new (non-economic) policy areas have been covered, such as security or infrastructure.

Regionalisation takes place within the complex context of the restructuring process of the governance and government architecture in the world. Regionalisation is taking place at the same time as globalization, decentralization and disintegration of nation states, and simultaneously with a diffusion of policy competences across levels and among actors (state and non-state), leaving us with a complex multi-level governance reality. In that reality, the macro-regional level is becoming a more important level, although in practically all policy areas, competences are being shared with one or more other levels. This, in turn, offers opportunities for functional synergies and/or tensions between levels.

Another important feature of the new regionalist wave is the fact that regionalization often implies multiple memberships (overlapping RIAs). In Africa, for example, according to recent UNECA data, only 7 countries are only belonging to only one integration arrangement; 27 countries belong to two arrangements, 18 countries to three arrangements, and one (DR Congo) even to four different regional integration arrangements (UNECA 2002, 2007).

Regional arrangements seem also to be more flexible, when positively viewed, or unstable, when negatively viewed, than national institutions. The world of regions is still constantly in motion. Recent events in South America (crisis in CAN, rapprochement between CAN and Mercosur, UNASUR, Venezuela's proposal for an alternative integration scheme, . . .), Africa (African Union, NEPAD[18], . . .), Asia (ASEAN+3, Eurasian Economic Community-Central Asian Economic Cooperation Organisation merger, . . .) and Europe (EU-25, negotiations with Croatia, Turkey, . . .) illustrate this very well.

A logical implication of the regionalisation process is the increased importance of interregional relations (interregionalism). More and more region-to-region relations are being established in different policy areas. The motivations behind interregionalism are diverse and are still in need of further analysis: the relative importance of political economy, economic, or bureaucratic forces is not clear. Domino effects are possible whereby, for example, donor regions like the EU impose the interregional model, thereby strengthening the regional counterparts, and therefore indirectly contributing to the development of region-to region interaction between third regions.

3.5. IMPLICATIONS OF NEW REGIONALISM FOR DEVELOPMENT FINANCE

What are the implications then of this (new) regionalization wave for development finance, and more specifically for the issues currently under debate as described above? The following paragraphs point to a number of these.

3.5.1 Aid Quantity, Volatility and Composition

Regionalization among donor countries and/or aid recipients is unlikely to be neutral in terms of the quantity, volatility and composition of aid flows.

Even if the overall quantity of aid would remain constant at the regional and global levels, aid levels at the national levels may be affected. If the movement from bilateral aid flows to region-to-region or asymmetrical patterns (country-to-region, region-to-country) is confirmed, one might expect the formation of a new geography of development finance and the gradual dissolution of post-colonial patterns of aid flows. As the European case illustrates well, the geographical distribution of regionally consolidated aid is not necessarily equal to the geographical distribution of the sum of pre-existing national aid flows.[19]

It is well possible that the tendencies as observed by Claessens et al. (2007) in their recent empirical work, are strengthened. These authors analysed the allocation of bilateral aid flows from 22 donor countries among 147 recipient countries, with data covering the 1970–2004 period. They found that since the 1990s the flows respond more to poverty levels, on the one hand, and institutional quality, on the other, and that variables like outstanding debt, country size and political and colonial linkages have lost terrain. However, they observe an important variation in aid selectivity levels of individual donors and attribute it to the particular institutional environments in the donor countries. This allows thinking that further shifts in the pattern of aid allocation in the same direction are to be expected when regional coordination becomes stronger.

Regional consolidation or coordination of aid on the donor side is also likely to have a beneficial effect on aid volatility, because it is equivalent to donor diversification from the perspective of the aid recipients. As we mentioned before, it has been empirically established that aid volatility is affected by the degree to which countries rely on a small number of individual donors (Fielding and Mavrotas, 2005, 2008).

The region-to-region interaction, implied by inter-regionalism, can be expected to open the door for innovative forms of aid. This is not only the case for macro-projects with a regional dimension, critically relying on scale, both on the donor and recipient sides, but also for micro-projects in border areas. Of particular importance appear to be the possibilities that are offered to donor countries to contribute to the financing of regional public goods (RPGs) (Cook and Sachs, 1999; Estevadeordal et al., 2004). Inter-regionalism also offers new possibilities in the area of AfT.

Although it is not always easy to clearly distinguish between different types of aid, in practice there seems indeed to be a tendency towards increased aid destined to the provision of public goods (incl. RPGs), compared to other (traditional) forms of foreign aid. The financing of global and regional public goods combined, as a percentage of total aid flows, grew for all OECD donors from 5 per cent in the beginning of the eighties to around 9 per cent at the end of the nineties (Arce and Sandler, 2002:14,81).

Infrastructure continues to stand out as one of the most important RPGs with great development potential. Recent innovative initiatives like the Initiative for the Integration of Regional Infrastructure in South America (IIRSA)[20] and NEPAD illustrate this (see below).

Regional security as an RPG and the interlinkages between regional cooperation/integration and security are also receiving growing attention both in political and in academic circles. On the ground, the African Union has illustrated this growing role for regional organizations in peacekeeping. On the basis of a review of different types of obstacles for regional cooperation towards different security problems, Stadtmüller (2005) argues that the effective involvement of regional organisations in conflict management depends on a series of factors, including: the commitment of the countries involved, shared values and interests, the institutional, financial and material capacity, and the legitimacy of the involvement. Although vital, these requirements are often missing in trouble spots in Africa and elsewhere and therefore often critically require external intervention. Graham and Felício (2006) presented a proposal for a regional-global security mechanism, which obviously also poses particular challenges for the financing of such an institutional construction.

The regionalization processes in the South are also likely to have an effect on investment flows; for example, through convergence of rules. And if regional integration implies monetary integration, exchange rate stability is expected to stimulate FDI (and trade) flows. Dollarization, like in Hong Kong, Ecuador or El Salvador, as a special case of monetary integration, is likely to have similar effects.

Finally, regionalization and regional coordination will not necessarily leave the political economy of aid flows unaltered. North-South asymmetries will not necessarily diminish and new forms of strategic donor behaviour are not to be excluded.

In the case of the EU, the most important donor in the world, regional coordination of aid facilitates the linkage between aid and trade, since trade policy is of exclusive regional competence.

3.5.2 Aid and the Financing of RPGs

The definition of RPGs evolved from purely technical definitions, inspired by Samuelson (1954), to more political and normative definitions (Kaul, 2003; Hettne and Söderbaum, 2004). Institutional economics also had its influence on the definition of public goods, including less visible and immaterial goods, like institutions and trust, in that category. Arce and Sandler's typology (based on two criteria: the degree of publicness of the goods and their aggregation technology), helps to understand for which classes of RPGs the value added of donor intervention is clearest and, therefore, for which RPGs donor intervention is most crucial and what the appropriate modality is. For example, in the case of an impure public good, the role of the donor consists in increasing access to the public good. In the case of club goods, its role is to subsidize access/entrance to the club for low-income countries (entities). For weakest-link RPGs, donors can raise the level of the RPG provision to a required minimum level.

Other arguments in favour of donor financing of RPGs include:

- the lack of regional leadership (linked to the free-rider problem)[21];
- the lack of a regional governance infrastructure and relatively weak regional economic linkages;
- regional tax collection politically and/or economically less or not being feasible;
- the positive contribution of RPGs to the provision of GPGs and possible positive spillovers for donors;
- economies of scale and/or scope (Arce and Sandler, 2002:32-).

Although donor intervention can be seen as a solution for the free-rider problem at the regional level, there should be awareness of the political aspects of it. Technical and economic arguments might clash here with principles of sovereignty.

3.5.3 The Financing of Infrastructure and the Role of Aid: The Case of NEPAD

Infrastructure is one of the eight priority sectors identified in NEPAD (others include education, health, agriculture, New Information & Communication Technologies, the environment, energy and access to the markets of developed countries (Wade, 2003). Needless to say, high-quality infrastructure is essential for attaining the MDGs. This is particularly true in the case of soft infrastructure (e.g. education, health and clean water supply and sanitation) which is strongly related to many of the MDGs. Improving substantially the volume as well as the quality of soft infrastructure in Africa will have significant implications for accelerating progress with many of the Goals in the region (Yuge, 2003).

The Short-Term Action Plan (STAP) for Infrastructure Development is also part of NEPAD's plans since NEPAD was created in 2001. More recently, the UN Millennium Project and the Commission for Africa reports recommended increased development partner support for infrastructure in the region. Along these lines the NEPAD Secretariat concludes that these developments are expected to lead to increased investment in the sector with profound implications for further growth in the region (Nkuhlu, 2005).

A series of World Bank reports on infrastructure seem to suggest that Africa is well behind other developing regions regarding infrastructure investment. The quality of infrastructure services is also very low in Africa compared to other regions (see WB reports; Obame, 2003). Africa's telecommunication infrastructure, for example, is the least developed in the world with less than 2 per cent of the world's telephone mainlines and just about 10 million telephones in 1999 (Mugabe, 2003; World Bank, 2000). The transport problem is also severe in the case of landlocked African economies in which freight expenses can absorb as much as 40 per cent of the value of traded goods compared with 4 per cent in the industrialized world (UNCTAD, 2001).

Needless to say, infrastructure development requires substantial resources in terms of aid flows and private capital. Aid remains an important source for infrastructure development although during the 1990s private capital flows increased substantially their share regarding the funding of infrastructure projects in the developing world. In the 1990s infrastructure for the developing world as a whole attracted substantial private capital flows reaching a peak of over US$120 billion in 1997 thus exceeding ODA flows to infrastructure that year. The Asian financial crisis of 1997, however, had a dramatic effect on private capital flows to infrastructure during the period 1997–1999 reducing them by almost 50 per cent during the above period. For the whole decade, however, total private investment in infrastructure (USD550 billion) was more than three times as much as total ODA flows (UNCTAD, 2001; PPI Database, 2000). Africa was not however the main recipient of the above spectacular increase in private sector financing of infrastructure since 80 per cent of private capital flows (in the form of FDI and portfolio flows) channeling to infrastructure sectors went to just six upper middle income countries – Brazil (receiving almost one third of the total), Argentina, Mexico, South Korea, Malaysia and Hungary. This means that ODA flows remain the principal source of financing infrastructure in developing countries and in particular in the Africa region (UNCTAD, 2001).

Promoting regional initiatives such as NEPAD for financing infrastructure development seems, thus, to be crucial. In this regard, linking up Africa with the global economy requires infrastructure investment that goes beyond the individual capacity of most African countries: regional or sub-regional approaches to the development of such infrastructure seem to be the right response in this case. Recent years have witnessed some progress in the region towards regional coordination in infrastructure projects: A Transport Protocol was signed recently in southern Africa to promote harmonization of transport policies between countries in the region; 23 states in West Africa agreed to liberalise air transport in the region breaking up the 40 year monopoly of *Air Afrique*. Also a project adopted under the auspices of the UNU Transport Decade (1990–2000) is currently linking nine trans-African highways with a view to connecting major cities in the region including Cairo to Cape Town on the north-south axis and Lagos to Mombassa on the east-west axis (UNCTAD, 2001). Furthermore, the NEPAD Secretariat is currently trying to secure funding for two or three key infrastructure projects for each of the five African regions. Possible candidates are a West African electric grid and power pool, regional gas pipelines and transport corridors (Herbert, 2003). Finally, NEPAD urgently needs to gain momentum by improving the quality of infrastructure as well as services and government management in Africa. This would involve *inter alia* the development of the necessary institutions for designing, implementing and monitoring infrastructure projects, adjusting trade and industrial policies and involving the local community, business and academia in the whole process.

3.5.4 Fiscal Response

Shifting aid flows from the national to the regional level, on the donor side and/or the recipient side, will not necessarily be neutral from a fiscal point-of-view. One should be aware that displacement or crowding out effects (see above) are possible, and the net effects in terms of fiscal efforts on the recipient side should be carefully monitored. This does not mean, however, that these effects will be necessarily negative. One could well put foreword the hypothesis according to which displacement effects might be reduced when aid is channeled through the regional level and ceases to compete directly with own national public expenditures of the recipient.

Following a similar reasoning, the provision of RPGs might well displace the provision of NPGs, especially if recipient countries are supposed to contribute to the RPG (the so-called neutrality theorem; Arce and Sandler, 2002:11). This is of particular relevance when, like in many cases, the (developmental) impact of RPGs depends on prerequisites, including the availability of NPGs. It has indeed been established that there is often a complementarity between regional and national public goods (Stålgren, 2000).

3.5.5 Institutions, Coordination and Absorptive Capacity

One uncertain factor in this discussion on the role of regional organizations in development finance is their absorptive capacity for aid. From a budgetary perspective, the regional level is indeed still very 'thin'[22], and the institutional strength of the regional bodies and the quality of regional policies, which are crucial in this respect as we have seen before, are very diverse, to say the least. This is connected to the more general observation of differing degrees of 'regionness' across regions. In some cases, shifting from the national to the regional level today will necessarily imply a weaker institutional capacity and weaker absorptive capacity. For the donors, timing will therefore be crucial when re-allocating aid among governance levels. On the other hand, the problem of institutional weakness is sometimes directly tackled by the aid programmes themselves; European regional programmes, for example, typically provide institutional support for the regional organizations, accompanying other aid flows.

Conditioning aid in terms of the quality of complementary policies in the host region, will be more problematic than in the case of bilateral aid relationships. Regional policies are not necessarily homogeneous and may well consist of a set of good and bad policies.

The overlap of different regional arrangements and organizations (see before), in combination with more institutional flexibility (instability?) will make it further difficult to establish a stable donor-recipient relationship.

The interregionalist model presents serious challenges to donor and recipient communities in terms of policy coordination. The EU-ACP case is a good example of the complex and multi-level institutional cooperation architecture that is emerging. Even when one focuses on governmental actors alone, a spaghetti-bowl of interactions (bilateral aid flows, country-to-region flows, region-to-country flows, EPA negotiations, Cotonou agreement, . . .) emerges between European actors (at the subnational, national and community levels), on the one hand, and ACP actors (at the national, regional and ACP levels), on the other. This situation has a cost in terms of policy coherence and aid effectiveness (Bretherton and Vogler, 1999; Söderbaum and Stålgren, 2009). This challenge of policy coordination has been recognized by the policy-makers at the European level (European Commission, 2005), although vested interests and historical patterns are serious obstacles to profound reforms.

In the case of RPGs, special attention should also be paid to the involvement of the recipient countries. These can be involved as a group or individually. A lot will again depend on the degree of integration that the group has reached and their choices and preferences in this matter; however, in some cases donor countries impose one or another modality, preferring either interregionalist or bilateral approaches. Strategic, technical and practical considerations all play a role in shaping policies.

Finally, the RPG framework has established that the problem of collective action is not identical in all policy areas. Therefore, regional cooperation in one particular field (with less resistance, more urgency or higher expected benefits) may trigger cooperation in other areas afterwards. Therefore, donor intervention may be crucial to achieve a critical mass or to sustain domino effects in terms of RPG provision. This coincides with the observed phenomenon of acquisition of autonomy by regional organizations (Arce and Sandler, 2002:34). Projects then become processes driven by permanent organizations.

3.5.6 Recipient Country Coordination: The Case of NEPAD

NEPAD is an example of a new initiative with respect to coordination of aid flows on the recipient side. NEPAD focuses on development aid and debt reduction as complementary external resources required in the short and medium term in order to fill the resource gap of US $64 billion so that the goal of reducing poverty by half in the region can be met by the year 2015. Needless to say, in the long run other sources of capital such as private capital flows but also domestic resource mobilization need to be addressed if the MDGs are to be met. With regard to development aid, the NEPAD Secretariat documents clearly state that ". . .. NEPAD will establish a forum of African countries so as to develop a common African position on Overseas Development Assistance (ODA) reform and to engage with the Development Assistance Committee (DAC) of the OECD and other donors in developing a charter underpinning the development partnership" (NEPAD Secretariat, 2005). NEPAD goes further by planning the following actions in this regard:

- Constitute an ODA forum for developing a common African position on ODA reform as a counterpart to the OECD-DAC structure.
- Engage, through the ODA forum, with donor agencies to establish a charter for the development partnership which would embody key principles for ODA reform.
- Support efforts of the Economic Commission for Africa to establish a PRSP Learning Group.
- Establish an independent mechanism for assessing donor and recipient country performance (NEPAD Secretariat, 2005).

These are obviously important developments in view of the need to accelerate progress regarding the MDGs in the region, at least as far as Goal 1 (see poverty reduction) is concerned. At the same time, they could strengthen further current efforts among donor agencies and the DAC to coordinate more efficiently aid flows to the region (see recent developments associated with the Paris Declaration on Aid Effectiveness in February 2005).

It has to be stressed though that the resource gap of US $64 billion referred to in the NEPAD document has caused a lot of discussion and debate recently. The figure comes from the World Bank and it is roughly calculated that to meet Goal 1 of halving the proportion of people living in poverty in the region by 2015 Africa will need to average an annual economic growth rate of 7 percent. Since the capital-output ratio is very low in the region, a high domestic investment rate of 33 per cent has to be assumed for this high growth rate to be achieved. However, taking into account that the domestic savings rate in Africa is 15 per cent and the FDI is 8 per cent (net) leaves a resource gap which is equivalent to 10 per cent of GDP i.e. USD64 billion in 2001. It has been argued that the above figure can be regarded as a dividing line between those wishing to use NEPAD to attract new financial resources to the region and those striving for an African-driven renaissance and for whom the above figure represents at most a political demand on the world's industrialized nations (Melchers, 2005).[23]

3.6. CONCLUDING REMARKS

The purpose of this chapter was to place the discussion on Aid for Trade in the context of the wider discussion on foreign aid flows. Particular attention was also paid to the implications of the growing importance of the regional governance level, both on the donor and on the recipient side.

Recent years have witnessed a changing environment for foreign aid and development finance in general and new issues have emerged partly due to the ongoing debate on how the Millennium Development Goals can be achieved by 2015 using existing sources of finance (predominantly foreign aid) but also other sources of development finance. The issues in the current debate on aid include *inter alia*, the question of the 'adequate' quantity and modality of aid, the need to address the volatility of aid, the (possible adverse) fiscal response to aid and the related issue of the overall macroeconomic management of aid increased aid flows in aid recipients, and – last but not least - the institutional requirements for aid effectiveness and the recipient countries' absorptive capacity.

At the same time, the new wave of regionalization in the world revealed *inter alia* new challenges for development aid in particular and development finance in general. On the one hand, regionalization can offer new opportunities for regional financial cooperation in the area of development assistance as has recently been reflected in new regional initiatives such as the New Partnership for Africa's Development (NEPAD); on the other hand, regional cooperation in the area of development finance can have profound implications for the existing foreign aid architecture.

Regionalisation has implications for both donors and aid recipients. A first implication is related to the need for more coordination among donors, on the one hand, and among aid recipients, on the other. A second implication is related to the fact that the whole development aid machinery will have to be adapted to an emerging architecture in which region-to-region relationships will be more important, at the cost of post-colonial ties. For individual aid recipients, this represents both challenges and opportunities. A third set of implications represent opportunities for aid policies to reduce volatility, to tackle development issues with a regional scale and/or cross-border characteristics, and to contribute to the provision of RPGs. Finally, the institutional preparedness of the recipient regions and the possible adverse fiscal responses at the national level should be carefully assessed.

NOTES

[1] Knack and Rahman (2004) define donor fragmentation as a large number of donors each with a small share of the total aid provided to a given recipient country.

[2] See also Roodman (2008) on the issue of donor proliferation and administrative burden in aid recipients, and Maxwell (2006) and Burall and Maxwell (2006) on the growing complexity of development aid 'architecture'.

[3] For a recent detailed discussion of the central issues related to development finance and development aid see Addison and Mavrotas (2008).

[4] Taken together, these altruistic donor initiatives seem to indicate that strategic donor behaviour where donors use their influence through foreign aid relations to achieve their own goals has declined. However, as argued quite convincingly by Burnell (2004), issues related to strategic donor behaviour and *realpolitik* have not lost importance in recent years. Moreover, strategic non-altruistic donor behaviour can be profitable (from the mighty donor's point of view) and can thus be expected to continue to dominate the aid allocation agenda in the near future if no safeguard is set in place (Mavrotas and Villanger, 2006).

[5] See Atkinson (2004) for an excellent discussion of new sources of development finance.

[6] Burnell (2004) provides a fascinating discussion of the changing landscape of aid in the 1990s; see also Addison et al. (2005a,b).

[7] See Collier and Dollar (2001, 2002), Beynon (2002, 2003), McGillivray (2003), Munro (2005) and Radelet (2008) on the selectivity issue.

[8] See Hansen and Tarp (2000, 2001), Dalgaard and Hansen (2001), Lensink and White (2001), Guillaumont and Chauvet (2001) and more recently Dayton-Johnson and Hoddinot (2003), Denkabe (2004), Chauvet and Guillaumont (2004), Dalgaard et al. (2004) and Ram (2004).

[9] On the crucial issue of the political economy of aid see the Special Issue edited by Sajal Lahiri and Katharina Michaelowa in the *Review of Development Economics* in 2006 and the papers therein, in particular regarding the role of institutions within aid agencies and the conflicting objectives adopted in many cases, the delivery mechanisms involved and the incentives which may affect the overall assessment of the various aid modalities adopted by donors (Lahiri and Michaelowa, 2006).

[10] See Bourguignon and Sundberg (2007a); Killick and Foster (2007); Collier (2006); de Renzio (2005); Gupta *et al.* (2005); Addison *et al.* (2005a,b); Heller (2005); Foster (2003); and Heller and Gupta (2002) among others.

[11] The term is attributed to White (1992).

[12] Empirical studies in the area of fungibility include Pack and Pack (1990, 1993), Khilji and Zampelli (1991) and Feyzioglu *et al.* (1998); see also White (1998) for a detailed discussion.

[13] Recently, aid volatility issues have also been briefly discussed within the context of aid to *difficult partnership countries* (Levin and Dollar, 2005).

[14] www.wto.org.

[15] For a recent comparative perspective on regional integration in different regional contexts, see different contributions in Farrell et al. (2005).

[16] Recent initiatives like the Central American Integration System (CAIS) and NEPAD are examples of this movement towards multidimensional forms of regional integration.

[17] See e.g. Sampson and Woolcock (2003).

[18] NEPAD's foundation document was adopted by the Assembly of African Heads of State and Government of the Organization of African Unity (now the African Union) in July 2001. The foundation document was subsequently ratified when the African Union was established in 2002.[18] NEPAD's key principles and messages include African ownership and responsibility for the development of the region, the promotion and advancement of democracy, human rights, good governance and accountable leadership, self-reliant development to reduce dependency on aid, building capacity in African institutions, promoting intra-Africa trade and investment, accelerating regional economic integration, advancing women, strengthening Africa's voice in international forums and forging partnerships with African civil society, the private sector, other African countries and the international community (Nkuhlu, 2005).

[19] On European development policies, see for example European Commission (2002, 2005) and Söderbaum and Stålgren (2006).

[20] See, www.iirsa.org.

[21] This role can also be played by regional advocacy networks, financed by donors.

[22] According to calculations by UNU-CRIS, the ECOWAS budget should be multiplied by 60 to reach a level comparable with the budget of the European Union (as a percentage of combined regional

GDP). The budget of the African Union should be multiplied by 10000... with most other regional organizations finding themselves somewhere in between these two extremes.

23 See also Luiz (2005) for an excellent recent discussion of issues related to NEPAD including a summary of the problems that may hinder the success of the plan.

REFERENCES

Addison, T. and G. Mavrotas (Eds.) (2008). *Development Finance in the Global Economy: The Road Ahead.* Basingstoke: Palgrave Macmillan for UNU-WIDER.

Addison, T., G. Mavrotas, and M. McGillivray (2005a). Aid, debt relief and new sources of finance for meeting the millennium development goals. *Journal of International Affairs*, 58(2), 113–127.

Addison, T., G. Mavrotas, and M. McGillivray (2005b). Aid to Africa: an unfinished agenda. *Journal of International Development*, 17(8), 989–1001.

J.E. Antipin, and G. Mavrotas (2006). On the empirics of aid and growth: a fresh look, UNU-WIDER Research Paper No. 2006/05.

Arce, D.G. and T. Sandler (2002). *Regional Public Goods: Typologies, Provision, Financing, and Development Assistance.* Stockholm: EGDI – Almkvist & Wiksell International.

Atkinson A.B. (Ed.) (2004).*New Sources for Development Finance.* Oxford: Oxford University Press for UNU-WIDER.

Baliamoune-Lutz, M. and G. Mavrotas (2008). Aid effectiveness: looking at the aid-social capital-growth nexus. *Review of Development Economics*, forthcoming.

Beynon, J. (2002). Policy Implications for Aid Allocations of recent research on aid effectiveness and selectivity. In: B. Mak Arvin (Ed.), *New Perspectives on Foreign Aid and Economic Development* (pp. 199–264). Westport, Connecticut: Praeger.

Beynon, J. (2003), Poverty Efficient Aid Allocations: Collier/Dollar Revisited, Economic and Statistics Analysis Unit Working Paper, Overseas Development Institute, London, (2).

Breslin, S., C.W. Hughes, N. Phillips and B. Rosamond (Eds.) (2002). *New Regionalisms in the Global Political Economy: Theories and Cases.* Routledge: London.

Bretherton, C. and J. Vogler (1999). *The European Union as a Global Actor.* Routledge: London and New York.

Bourguignon, F. and M. Sundberg (2007a). Absorptive Capacity and Achieving the Millennium Development Goals. In Mavrotas, G. and A. Shorrocks (Eds.), *Advancing Development: Core Themes in Global Economics* (pp. 640–663), Basingstoke: Palgrave Macmillan for UNU-WIDER.

Bourguignon, F. and M. Sundberg (2007b). Is foreign aid helping? Aid effectiveness – opening the black box. *American Economic Review*, (97)2, 316–321.

Bulíř, A. and J. Hamann (2003). Aid Volatility: An Empirical Assessment, IMF Staff Papers No. 5.

Burall, S. and S. Maxwell (2006). Reforming the International Aid Architecture: Options and Way Forward, Overseas Development Institute, ODI Working Paper No. 278, October.

Burnell, P. (2004). Foreign aid resurgent: New spirit or old hangover, Research Paper No.2004/44, World Institute for Development Economics Research, Helsinki.

Burnside, C. and D. Dollar (2000). Aid, Policies and Growth. *American Economic Review*, 90(4), 847–868.

Cassen, R. (1994). *Does Aid Work?.* Oxford: Oxford University Press.

Chauvet, L. and P. Guillaumont (2004). Aid effectiveness in an unstable environment, Mimeo, CERDI, France.

Chauvet, L. and P. Guillaumont (2008). Aid, volatility and growth again: when aid volatility matters and when it does not. *Review of Development Economics*, forthcoming.

Claessens, S., D. Cassimon and B. Van Campenhout (2007). Empirical Evidence on the New International Aid Architecture, IMF Working Paper No. WP/07/277.

Clemens, M., S. Radelet and R. Bhavnani (2004). Counting Chickens when they Hatch: The Short-term Effect of Aid on Growth, Working Paper No. 44, Washington, DC: Center for Global Development.

Collier, P. (2006). *What Can We Expect from More Aid to Africa?*, Unpublished.

Collier, P. and D. Dollar (2001). Can the world cut poverty in half? How policy reform and effective aid can meet international development goals. *World Development*, 29(11), 1787–1802.
Collier, P. and D. Dollar (2002). Aid allocation and poverty reduction. *European Economic Review*, 45(8), 1470–1500.
Collier, P. and D. Dollar (2004). Development effectiveness: what have we learnt? *Economic Journal*, 114(496), F244–F271.
Cook, L.D. and J. Sachs (1999). Regional Public Goods in International Assistance. In I. Kaul, I. Grunberg and M.A. Stern (Eds), *Global Public Goods. International Cooperation in the 21st Century* (pp. 436–449), New York: Oxford University Press.
Cuyvers, L., P. De Lombaerde, E. De Souza and D. Fielding (2005). Regional monetary cooperation and integration. In: M. Farrell, B. Hettne and L. Van Langenhove (Eds), *Global Politics of Regionalism. Theory and Practice* (pp. 120–136), London: Pluto Books.
Dalgaard, C.-J. and H. Hansen (2001). On aid, growth and good policies. *Journal of Development Studies*, 37(6), 17–41.
Dalgaard, C.-J., H. Hansen and F. Tarp (2004). On the empirics of foreign aid and growth. *Economic Journal*, 114(496), F191–F216.
Dayton-Johnson, J. and J. Hoddinot (2003). *Aid, policies and growth redux*, unpublished manuscript.
De Lombaerde, P. (2003). Book review of 'new regionalism in the global political economy. *Journal of Common Market Studies*, 41(5), 968–969.
Denkabe, P. (2004). Policy, Aid, and Growth: A Threshold Hypothesis. *Journal of African Finance and Economic Development*, 6, 1–21.
de Renzio, P. (2005). Can More Aid be Spent in Africa?, *ODI Opinions*, (30). London: Overseas Development Institute.
Doucouliagos, H. and M. Paldam (2005). The Aid Effectiveness Literature: The Sad Results of 40 Years of Research, Department of Economics Working Paper 2005.05, University of Aarhus.
Easterly, W. (2006). *The White Man's Burden: Why the West's Efforts to Aid the Rest Have Done So Much Ill and So Little Good*. New York: Penguin Press.
Easterly, W. (2007). Was Development Assistance A Mistake? *American Economic Review*, 97(2), 328–332.
Easterly, W., R. Levine and D. Roodman (2004). New data, new doubts: a comment on burnside and dollar's 'aid, polices and growth (2000). *American Economic Review*, 94(3), 774–780.
Estevadeordal, A., B. Frantz and T.R. Nguyen (Eds) (2004). Regional Public Goods, Washington: IDB-ADB.
European Commission (2002). Annual Report 2001 from the Commission to the Council and the European Parliament: On the EC Development Policy and the Implementation of the External Assistance, Brussels: Commission of the European Communities.
European Commission (2005). Proposal for a Joint Declaration by the Council, the European Parliament and the Commission on the European Union development Policy, The European Consensus, COM(2005)311Final, Brussels: Commission of the European Communities.
Farrell, M. (2005). The Global Politics of Regionalism – An Introduction. In: M. Farrell, B. Hettne and L. Van Langenhove (Eds.), *Global Politics of Regionalism. Theory and Practice* (pp. 1–17). London: Pluto Books.
Farrell, M., B. Hettne and L. Van Langenhove (Eds.) (2005), *Global Politics of Regionalism. Theory and Practice*, London: Pluto Books.
Fawcett, L. and A. Hurrell (Ed.) (1995). *Regionalism in World Politics: Regional Organization and International Order*. Oxford and New York: Oxford University Press.
Feyzioglu, T., V. Swaroop and M. Zhu (1998). A panel data analysis of the fungibility of aid. *World Bank Economic Review*, 12(1), 29–58.
Fielding, D. and G. Mavrotas (2005). The Volatility of Aid, WIDER Discussion Paper, World Institute for Development Economics Research, 2005/2006.
Fielding, D. and G. Mavrotas (2008). Aid volatility and donor-recipient characteristics in 'difficult partnership countries', *Economica*, 75(299), 481–494.

Foster, M. (2003). The Case for Increased Aid, Report to the Department for International Development (DFID), London.

Gavin, B. and P. De Lombaerde (2005). Economic Theories of Regional Integration. In Farrell, M., B. Hettne and L. Van Langenhove (Eds.), *Global Politics of Regionalism. Theory and Practice* (pp. 69–83), London: Pluto Books.

Graham, K. and T. Felício (2006). Regional Security and Global Governance: A Proposal for a Regional-Global Security Mechanism, Egmont Papers, Brussels: Royal Institute for International Relations.

Guillaumont, P. and S. Guillaumont-Jeanneney (2007). Big Push versus Absorptive Capacity: How to Reconcile the Two Approaches, UNU-WIDER Discussion Paper 2007/05, World Institute for Development Economics Research, Heslinki.

Gupta, S., R. Powell and Y. Yang (2005). The Macroeconomic Challenges of Scaling Up Aid to Africa, IMF Working Paper 05/179, International Monetary Fund, Washington, DC.

Hansen, H., and F. Tarp (2000). Aid Effectiveness Disputed. *Journal of International Development*, 12(3), 375–398.

Hansen, H. and F. Tarp (2001). Aid and Growth Regressions. *Journal of Development Economics*, 64(2), 547–570.

Heller, P.S. (1975). A Model of Public Fiscal Behaviour in Developing Countries; Aid, Investment and Taxation. *American Economic Review*, 65(3), 429–445.

Heller, P.S. (2005). Pity the Finance Minister: Issues in Managing a Substantial Scaling Up of Aid Flows, IMF Working Paper *05/180*.

Heller, P. and S. Gupta (2002). More Aid – Making It Work for the Poor. *World Economics*, 3(4), 131–146.

Herbert, R. (2003). Implementing NEPAD: A Critical Assessment. In *Africa Report: Assessing the New Partnership*, Ottawa, Canada: North-South Institute.

Hettne, B., A. Inotai and O. Sunkel (1999–2001). *The New Regionalism Series* (Vol. I–V). London: Macmillan Press.

Hettne, B. and F. Söderbaum (2000). Theorising the Rise of Regionness. *New Political Economy*, 5(4), 457–472.

Hettne, B. and F. Söderbaum (2004). *Regional Cooperation. A Tool for Addressing Regional and Global Challenges*, International Task Force on Global Public Goods, Göteborg, 9.

Hudson, J. (2004). Introduction: Aid and Development. *Economic Journal*, 114(496), F185–F190.

Hudson, J. and P. Mosley (2006). Aid Volatility, Policy and Development, Paper presented at the WIDER Development Conference on Aid: Principles, Policies and Performance, WIDER Helsinki, 16–17 June.

Kaul, I. (2003). Re-Defining Public Goods: Why, How, and to What Effect, Paper prepared for the *Conference on Regional Public Goods*, UNU-CRIS, Bruges, 20–21 November.

Khilji, N., and E. Zampelli (1991). The Fungibility of US Assistance to Developing Countries and the Impacts on Recipient Expenditures: a Case Study of Pakistan. *World Development*, 19(8), 1095–1106.

Killick, T. and M. Foster (2007). The Macroeconomics of Doubling Aid to Africa and the Centrality of the Supply Side. *Development Policy Review*, 25(2), 167–192.

Knack, S. and A. Rahman (2004). Donor Fragmentation and Bureaucratic Quality in Aid Recipients, World Bank Policy Research Working Paper 3186.

Lahiri, S. (Ed.) (2007). *Theory and Practice of Foreign Aid*, Amsterdam/Oxford: Elsevier.

Lahiri, S. and K. Michaelowa (2006). The Political Economy of Aid, Special Issue. *Review of Development Economics*, 10(2), 177–344.

Lensink, R. and H. White (2001). Are there negative returns to aid?. *Journal of Development Studies*, 37(6), 42–65.

Levin, V. and D. Dollar (2005). The Forgotten States: Aid Volumes and Volatility in Difficult Partnership Countries, Paper prepared for DAC Learning and Advisory Process on Difficult Partnerships.

Luiz, J. (2005). The New Partnership for African Development: Questions regarding Africa's response to its underdevelopment. *Journal of International Development*, forthcoming.

Mavrotas, G. (2002a). Foreign Aid and Fiscal Response: Does Aid Disaggregation Matter? *Weltwirtschaftliches Archiv* (*Review of World Economics*), 138(3), 534–559.

Mavrotas, G. (2002b). Aid and Growth in India: Some Evidence from Disaggregated Aid Data. *South Asia Economic Journal*, 3(1), 19–49.
Mavrotas, G. (2005). Aid Heterogeneity: Looking at Aid Effectiveness from a Different Angle. *Journal of International Development,* 17(8), 1019–1036.
Mavrotas, G. and P. Nunnenkamp (2007). Foreign Aid Heterogeneity: Issues and Agenda. *Review of World Economics*, 143(4), 585–595.
Mavrotas, G. and B. Ouattara (2006a). Public Fiscal Behaviour and Aid Heterogeneity in Aid-recipient Economies. *Journal of Developing Areas*, 39(2), 1–15.
Mavrotas, G. and B. Ouattara (2006b). Aid disaggregation, endogenous aid and the public sector in aid-recipient economies. *Review of Development Economics*, 10(3), 432–449.
Mavrotas, G. and H. Reisen (2007). The Multilateral Development Finance Non-System, Paper presented at the Experts' Workshop on Performance and Coherence in Multilateral Development Finance, Berlin, 29–30 January.
Mavrotas, G. and E. Villanger (2006). Multilateral aid agencies and strategic donor behaviour, UNU-WIDER Discussion Paper No. 2006/02.
Maxwell, S. (2002), More Aid? Yes – and Use it to Reshape Aid Architecture, *ODI Opinions*, London: ODI.
Maxwell, S. (2006). Aid Architecture: A Blueprint for the Future, ODI Annual Report 2006, London: ODI.
McGillivray, M. (2003). Aid effectiveness and selectivity: integrating multiple objectives in aid allocations, Paper presented at the Joint OECD DAC/Development Centre Aid Seminar, Paris: OECD Headquarters.
McGillivray, M., S. Feeny, N. Hermes and R. Lensink (2006). It Works; It Doesn't; It Can, But That Depends. 50 Years of Controversy over the Macroeconomic Impact of Development Aid. *Journal of International Development*, 18(7), 1031–1050.
Melchers, K. (2005). New Partnership for African Development: A reform Agenda for Africa – Initial Success and Criticism. *Agriculture and Rural Development*, 12(2).
Mosley, P. (1987). *Overseas Aid: Its Defense and Reform*. Brighton: Wheatsheaf Press.
Mugabe, J. (2003). The Keys to Africa's Sustainable Development: Science, Technology and Investment. In *Africa Report: Assessing the New Partnership*. Ottawa, Canada: North-South Institute.
Munro, L. (2005). Focus-Pocus? Thinking critically about whether aid organizations should do fewer things in fewer countries. *Development and Change*, 36(3), 425–447.
NEPAD Secretariat (2005), "The Capital Flows Initiative", Johannesburg: NEPAD.
Nkuhlu, W.L. (2005). The New Partnership for African Development: The Journey So Far, Johannesburg: NEPAD Secretariat.
Obame, J.-C. (2003). Overview of Investment and Financing to Africa: Challenges and Risk Management, Paper presented at the Africa Day Symposium, UNU, Tokyo, 13 May.
Ouattara, B. (2006). Foreign aid, public savings displacement and aid dependency in Cote d"Ivoire. *Oxford Development Studies*, 35(1), 33–46.
Pack, H., and J.R. Pack (1990). Is Foreign Aid Fungible?: The Case of Indonesia. *Economic Journal*, 100(399), 188–194.
Pack, H. and J.R. Pack (1993). Foreign Aid and the Question of Fungibility.*Review of Economics and Statistics*, 75(2), 258–265.
Pallage, S. and M. Robe (2001). Foreign Aid and the Business Cycle. *Review of International Economics*, 9(4), 636–667.
Radelet, S. (2006). A Primer on Foreign Aid, Working Paper No.92, Center for Global Development, Washington D.C.
Radelet, S. (2008). The Millennium Challenge Account: Transforming US Foreign Assistance Policy? In T. Addison and G. Mavrotas (Eds.), *Development Finance in the Global Economy: The Road Ahead*. Basingstoke: Palgrave Macmillan for UNU-WIDER.
Ram, R. (2004). Recipient Country's 'Policies' and the Effect of Foreign Aid on Economic Growth in Developing Countries: Additional Evidence. *Journal of International Development*, 16(2), 201–211.
Riddell, R. (2007). *Does Foreign Aid Really Work?* Oxford: Oxford University Press.

Roodman, D. (2008). Aid Project Proliferation and Absorptive Capacity. In G. Mavrotas and M. McGillivray (Eds.), *Development Aid: A Fresh Look*. Basingstoke: Palgrave Macmillan for UNU-WIDER (forthcoming).

Sampson, G. and S. Woolcock (2003). *Regionalism, Multilateralism, and Economic Integration. The Recent Experience*. Tokyo: UNU Press.

Samuelson, P.A. (1954). The Pure Theory of Public Expenditure. *Review of Economics and Statistics*, 36(4), 387–389.

Stadtmüller, E. (2005). Regional Dimensions of Security. In M. Farrell, B. Hettne and L. Van Langenhove (Eds), *Global Politics of Regionalism. Theory and Practice* (pp. 104–119). London: Pluto Books.

Stålgren, P. (2000). Regional Public Goods and the Future of International Development Co-operation. A Review of the Literature on Regional Public Goods, EGDI Working Paper, No. 2.

Söderbaum, F. and P. Stålgren (Eds) (2009). *The European Union and the Global South*. Boulder/London: Lynne Riener Publishers (forthcoming).

Tarp, F. (Ed.) (2000). *Foreign Aid and Development: Lessons Learnt and Directions for the Future*. London: Routledge Studies in Development Economics.

Thiele, R., Nunnenkamp, P. and Dreher, A. (2007). Do donors target aid in line with the millennium development goals? A sector perspective of aid allocation. *Review of World Economics*, 143(4), 596–630.

UNCTAD (2001). Third United Nations Conference on the Least Developed Countries (LDCIII) – Infrastructure Development Session, Brussels, 19 May.

UNECA (2002). Annual Report on Integration in Africa 2002, UN Economic Commission for Africa, Addis Ababa.

UNECA (2007). Regional Economic Integration in Africa. In P. De Lombaerde (Ed.), *Multilateralism, Regionalism and Bilateralism in Trade and Investment. 2006 World Report on Regional Integration* (pp. 127–157), Dordrecht: Springer

Wade, A. (2003). The Role of Infrastructures in Development and Regional Integration, Paper presented at the Africa Day Symposium, UNU, Tokyo, 13 May.

White, H. (1992). The macroeconomic impact of development aid: A critical survey. *Journal of Development Studies*, 28(2), 163–240.

White, H. (2008). Evaluating Aid Impact. In G. Mavrotas and M. McGillivray (Eds.), *Development Aid: A Fresh Look*. Basingstoke: Palgrave Macmillan for UNU-WIDER (forthcoming).

World Bank (1998). *Assessing Aid: What Works, What Doesn't and Why*. Oxford University Press: Oxford.

World Bank (2000). *Can Africa Claim the 21st Century?* Washington D.C. http://go.worldbank.org/Z7KUASD0N0

World Bank (2007). Aid Architecture: An Overview of the Main Trends in Official Development Assistance Flows. Washington: World Bank.

Yuge, A. (2003). Remarks at the Panel Discussion on Financing Infrastructure Projects in Developing Countries and in Africa: Challenges and Implementation, Africa Day Symposium, UNU, Tokyo, 13 May.

SANOUSSI BILAL AND FRANCESCO RAMPA, ECDPM

CHAPTER 4

WHAT DOES THE EUROPEAN EXPERIENCE TELL US ON AID FOR TRADE?

Since the launch of the aid-for-trade (AfT) initiative at the World Trade Organization (WTO) Ministerial Conference in Hong Kong in December 2005, the debate on the developing countries need for support to benefit from trade(-related) reforms and liberalisation has attracted much attention and generated several new policy initiatives. Yet, development assistance for trade matters is nothing new.[1] Experience by multilateral agencies, as well as regional and bilateral donors can prove most useful in better understanding what has worked and what has been less effective.[2] In this regard, the experience of the European Union (EU), as the largest donor in the world and a main provider of aid for trade, is of great interest.

The purpose of this Chapter is to review that European experience, in particular in a regional and international context, identifying some key lessons notably from the cooperation between the EU and the Africa, Caribbean and Pacific (ACP) countries, one of the prime examples of a North-South partnership with a strong trade and development dimension.[3]

4.1. AID FOR TRADE AND THE EU

Aid for trade has been and will continue to be an important dimension of the EU development assistance to developing countries.

4.1.1 New EU Strategy on Aid for Trade

The new approach by the EU on aid for trade is outlined in a policy document, the *EU Strategy on Aid For Trade*, adopted on 15 October 2007. The Strategy reaffirms the commitment by the EU (EU member states and European Commission) to increase its annual spending on trade-related assistance to €2 billion each year by 2010.[4] Perhaps more importantly, it provides a policy framework which builds on important principles like enhancing the pro-poor focus and quality of aid-for-trade, as well as increasing the EU's own capacity to provide such aid. Given that its objectives and guidelines apply to both EU member states and to the European Commission, such a framework provides an opportunity to forge a common and integrated EU approach to aid-for-trade implementation. Moreover, the Strategy is based on a demand-driven approach, aimed at "delivering an effective response to countries' own trade-related

P. De Lombaerde, L. Puri (eds.), *Aid for Trade: Global and Regional Perspectives*,
United Nations University Series on Regionalism 2,
DOI 10.1007/978-1-4020-9455-2_5,

priorities", and builds on existing EU commitments on policy coherence, enhanced division of labour and the *Paris Declaration on Aid Effectiveness*.[5]

The new EU Strategy must be seen in the context of the Doha Round and the EPAs negotiations, as it was agreed one month before the WTO's 'global review' on aid-for-trade in Geneva on 20-21 November 2007 and a couple of months before the initial end of 2007 official deadline for the conclusion of trade negotiations between the ACP and the EU on new economic partnership agreements (EPAs). The Strategy also aims at paving the way to a more concrete aid for trade action plan, to be elaborated in 2008, and thus as a mean to show that the EU has serious and significant aid for trade ambitions.

Following the launch of the aid-for-trade initiative at the WTO Ministerial Conference in Hong Kong in December 2005, expectations were high. However, uncertainty quickly set in over the thematic and geographical scope of the plan, as well as the financial resources available to execute it. What emerged from the recent WTO 'global review' , mainly a pledging exercise for donors, was that such an initiative in the multilateral arena will not comprise a new aid envelope or a specific delivery mechanism. The WTO will most likely remain a simple forum for monitoring the pay-out of global commitments on aid-for-trade, with different donor and recipient countries opting for a range of programmes and delivery mechanisms.

Given the modest outcome of the 'global review' and the continued difficulties to substantiate the WTO process on aid-for-trade, the implementation of the EU Strategy can thus be regarded as a concrete test for whether aid-for-trade is going to be 'business as usual' or an innovative instrument for more effective trade-related assistance. This is particularly relevant for those ACP countries that have or will sign an EPA, since the Strategy will have to respond immediately to specific trade-related needs – although aid-for-trade is not, and should not, be conditional on the conclusion of an EPA.[6]

4.1.2 Level of Support

But the EU is already doing a lot in terms of aid for trade for developing countries. The EU is among the main driving forces of AfT programmes and a substantial donor, at national, regional and multilateral levels. Aid for trade features throughout the European Community development cooperation agenda and over recent years several billion has been spent on trade capacity building and related programmes[7], a support which has been deemed as generally adequate to the beneficiaries' absorptive capacity.[8]

It is important to note that the determination of actual levels of funds available for AfT support crucially depends on the definition of the scope of the assistance considered as 'trade-related' and the effective classification of the aid accordingly. The WTO Aid for Trade Task Force recommended a broad definition of the scope of AfT interventions, comprising the following six categories, which have been increasingly used by donors and in all AfT debates:

1. support for trade policy and regulations;
2. trade development;
3. trade-related infrastructure;
4. building productive capacity;
5. trade-related adjustment;
6. other trade-related needs.

In recent years, most donors have concentrated their aid for trade efforts in the first two sub-categories only, classified in the Joint WTO/OECD Database as trade-related technical assistance and capacity building (TRTA/CB).[9] Inclusion of the other categories in the new AfT debates has created uncertainty in the donor community on how exactly to distinguish between them and report the different assistance activities.

The €2 billion EU commitment on AfT covers only the first two categories, support for trade policy and regulation and trade development. The EU has recognized the importance and need to cover the other four categories as well[10], but has not pledge any specific amount of support. Yet, these are the categories of AfT with the most crucial needs from developing countries and where the EU is currently spending most of its aid for trade.

Half of the annual €2 billion EU commitments by 2010 on trade-related assistance should come from EU member state aid budgets and the other half from the European Commission. The Commission is on safe ground, spending already close to €1 billion per year on AfT (€960 million in 2006), in part due to the agreed 2008–2013 10th European Development Fund (EDF) for ACP countries. It is thus the EU member states which are in the spotlight, having committed only € 370 million on trade-related assistance in 2006. Some EU member states have a long way to go before the collective pledge to increase their trade related aid to €1 billion annually by 2010.

4.1.3 Aid for Trade Activities

In terms of type of support, AfT by the European Community covers a large range of activities at the national level, the regional level and, in the case of the EDF, at an all-ACP level. Projects target implementation of international trade commitments, preparation for negotiations, institutional support such as for trade-policy formulation, private-sector development support, infrastructure improvements and product safety schemes. The European Community therefore already provides funds and programmes in all of the abovementioned AfT categories, but with two interrelated downsides.

The first is that the AfT activities for one region or country in most cases are not designed in a holistic manner and under a coherent framework. This emerges from external evaluations of European Community activities, according to which the scope often is not properly defined under each AfT initiative or cluster of interventions (ADE, 2004). This leads to:

(1) Short-term needs, like improving preparation for negotiations, are generally satisfied, especially at the regional level, but systemic reforms of trade policy and institutions were not induced, especially at the national level.
(2) Support to the private sector is generally provided horizontally across sectors of activity, without differentiating actions according to whether the supported industry or enterprise belongs to an import-competing or export industry.
(3) The main incentives for using the services offered under some projects were their low cost or the fact that they provided concession schemes by which to bypass the constraints of the business environment instead of addressing them.

The second downside reflects the definitions of aid for trade adopted by the European Community itself. The European Community is the largest single donor in the world for trade policy and regulations and trade development programmes (categories 1 and 2 of AfT), but other forms of support, for instance funds for adjustments to changing market conditions (such as in the context of the EU sugar market reform) or for infrastructure development (such as the EU-Africa Infrastructure Partnership Trust Fund) are not classified by the European Community as trade-related assistance (European Commission, 2006). Moreover, although 'general budget support' provided by the European Community (including to help offset government revenue loss due to liberalisation) is on the increase (including as a share of EDF), it is very difficult to break it down to distinguish the trade-related support elements. These problems of classification and reporting[11] make it difficult to assess the exact scope and amounts of aid for trade already provided by the European Community and the levels actually available in the framework of new initiatives and pledges.

The current debate in Europe on AfT and especially on the joint EU AfT strategy could be helpful to clarify operational issues related to definition of the scope and classification of AfT. It should be a useful contribution to the multilateral debates as well. This is particularly important as scope, definitions and effective classification of AfT activities will have consequences for the debate on the 'additionality' of aid for trade. Indeed, it is by no means certain that current pledges will actually lead to any additional AfT. Not only it is unclear how additionality is to be measured, but also the amounts announced differ and have been re-announced several times, with a risk of simple re-labelling of existing aid activities as AfT, with no fresh support for trade-related matters.[12]

Despite such downsides, one particularly positive aspect that is usually associated with the European Community support is that the scope of assistance covers also the regional level of intervention, going beyond the national dimension. The WTO Aid for Trade Task Force recognised that "many countries require cross-border infrastructure and regional policy cooperation to trade more effectively". However, the lack of regional AfT has been identified as one of the major issues not yet adequately addressed in current multilateral discussions on AfT (ILEAP, 2007). The European Community is well advanced on tackling this regional dimension. Regional-level assistance features prominently in both in the programming of the European Development Fund in general and in particular trade-related interventions. Trade is always a central part of the assistance offered by the EC for regional integration, based on its open regionalism policy and focusing on support for regional integration organisations, through which AfT interventions are identified and channelled.[13] In the Regional Integration Programmes (RIPs) of some ACP regions, assistance related to trade and trade policy dominates the cooperation agenda, with AfT commitments comprising over half of total funding. However, so far the European Community assistance has failed to effectively support mainstreaming trade into national development strategies through National Indicative Programmes (NIPs).[14]

4.1.4 Lessons on the Levels and Scope of AfT

A number of lessons can be drawn from the levels and scope of aid for trade in Europe. First, the recognition that trade can make a vital contribution to development should be matched by a policy decision by donors to accompany trade reforms in developing countries with substantial levels of funding for programmes to help them benefit more from trade.[15] The EU has identified trade as one of the priorities of its development policy, but this would not be enough if the EU member states, and particularly the European Community, were not to mobilise substantial financial resources in this direction. Other donors would probably have to step up their trade-related programmes, in addition to their existing development assistance, to make the AfT initiative effective and credible and prevent it from conflicting with other developmental needs and funding allocations.

A second, related dimension is that definition of the scope of aid for trade programmes should be broad, taking into account the vast range of needs to be met (for several stakeholders) to effectively transform trade opportunities into development realities. The experience of the European Community on aid for trade suggests that the broader the scope of intervention, the more important proper articulation, definition and reporting of AfT initiatives will be. To be effective, all aid for trade activities within a single country should be designed in a holistic manner and under a coherent framework. This also reveals the key role of coordination among different implementing agencies and actors, including harmonisation of practices, procedures and requirements of various donors. Moreover classification and reporting methodologies for AfT interventions should be clarified in detail in the early stages of programming assistance, to avoid confusion on the exact scope and amounts of the AfT actually available and any risk of a re-labelling of existing support.

In terms of geographical scope, another specific lesson from the European experience is that AfT initiatives should encompass also region-wide programmes, which can be especially successful in accompanying regional integration. At the same time, this should not mean losing focus on the national level.[16] Involvement of institutions coordinating regional integration in preparation of the AfT national-level strategies would facilitate the identification of the trade-related support needed by each member country to implement the regional integration reforms and commitments. The European Community experience provides a prime example of an assistance framework with specific regional mechanisms for delivery of support (e.g. the RIPs for the ACP) in which an attempt is made to coordinate the work of national and regional institutions (see Box 4.1.). Improvements are needed though to strengthen the linkages and complementarity between the regional and country levels of interventions and distinguish national and regional needs.

4.2. THE COTONOU AGREEMENT FRAMEWORK: AN EVOLVING AID AND TRADE PARTNERSHIP

The relations between the EU and the African, Caribbean and Pacific countries, governed since 1975 by successive Lomé Conventions and since 2000 the Cotonou Partnership Agreement (CPA), have always been a comprehensive partnership which have covered and integrated both trade and development concerns. It is thus a prime example of a North-South relationship covering national and regional aid for trade schemes. Indeed, the ACP and EU were among the first developed and developing country partners to establish tight linkages between trade and development as exemplified by Article 34 of the CPA:

> Economic and trade cooperation shall aim at fostering the smooth and gradual integration of the ACP States into the world economy, with due regard for their political choices and development priorities, thereby promoting their sustainable development and contributing to poverty eradication in the ACP countries.

The link between trade and other policies (accompanying measures) has become even more central with the negotiations and conclusion of comprehensive free trade agreements between the EU and ACP regions, called economic partnership agreements. Article 37.3 of the CPA states that in the context of the shift to such a new trading arrangement:

> The preparatory period shall also be used for capacity-building in the public and private sectors of ACP countries, including measures to enhance competitiveness, for strengthening of regional organisations and for support to regional trade integration initiatives, where appropriate with assistance to budgetary adjustment and fiscal reform, as well as for infrastructure upgrading and development, and for investment promotion.

This is an explicit recognition that without adequate policies and resources to adjust to economic transformation and to produce and market their goods competitively (including through cooperation with development partners), ACP countries are unlikely to fully benefit from international economic integration. The Aid for Trade initiative builds on an exactly the same premises.

Under the overall framework of the Cotonou Agreement, a large share of European assistance to ACP countries is channelled through the European Development Fund. The European Commission administers the EDF together with ACP countries and on behalf of the EU member states, also overseeing disbursement of European Community aid, including trade-related development support. EDF funds are provided in the form of grants and technical assistance disbursed in five-year cycles through multi-annual programming in national and regional indicative programmes (NIPS/RIPs) as well as all-ACP financial envelopes (see Box 4.1.). In addition, the EU member states have their own bilateral aid programmes (including aid for trade) and implementation agencies that are not financed by the EDF.

An important positive aspect of the EDF is that it is framed in the overall partnership spirit between the ACP and EU, whereby aid should not simply flow from donors to recipients, but should be an instrument to achieve common objectives, resulting from joint programming and through joint management. Moreover, the EDF management should be guided by key underlying principles and characteristics of effective donor assistance, many of which have since been emphasised by recommendation of the WTO Aid for Trade Force: long-term character; predictability; accessibility; demand driven; flexibility and broad range of instruments (i.e. national, regional, sectoral coverage); partnership; light conditionality; joint management by donor and recipient; and country ownership.

Unfortunately, these good principles are not always effectively translated into practice, as discussed below.[17]

An overarching principle is that for trade and aid to work together in delivering on development goals there has to be a shared vision that trade growth and development cooperation should accompany each other to make a vital contribution to poverty reduction. This is to some extent the case for the ACP and the EU, as together they have chosen aid and trade as two of the pillars of their partnership (enshrined in the CPA). But this cannot be taken as granted for all donor and recipient countries. Such a shared vision entails a partnership approach whereby the parties choose aid and trade as instruments to achieve common objectives and do not consider aid as compensation by the donor for trade liberalisation by the recipient country. Though this latter ‘stick and carrot’ approach should be unacceptable from both the donor and recipient perspectives, many stakeholders nonetheless indicate this is a concrete risk with aid for trade initiatives (Oxfam, 2005).

4.3. ASSESSING THE EU AFT: OWNERSHIP, CAPACITY AND MONITORING

The governance structure and the performance focus (or lack of it) of aid are two important elements of a development assistance framework. They significantly affect the quality of aid provided by the donor (supply side), the ownership and capacity of the recipient (demand side), as well as the ability to coordinate and monitor aid

Box 4.1. EDF programming instruments

Country and Regional Strategy Papers (CSPs and RSPs) are the main strategic frameworks for the programming of EU assistance and offer guidelines for allocation and implementation of EDF funds in different intervention areas. These papers set up the political guidelines for the implementation of cooperation policies and are instruments for directing, managing and reviewing EC assistance programmes, including on trade. The funds attached to the CSPs and RSPs are disbursed through multi-annual programming (in five-year cycles) in national and regional indicative programmes (NIPs and RIPs). The ACP authorities responsible for NIP and RIP implementation are the National and Regional Authorising Officers (NAOs and RAOs). In many national or regional assistance programmes, considerable funds have been explicitly reserved for aid for trade, and increasingly so in relation to EPAs.

Aid for trade instruments accessible to all ACP countries and regional institutions illustrate the scope of the European Community AfT, from preparation of negotiations, to longer term capacity building, to covering needs specific to an economic sector or a trade-related constraint. Under the 9th EDF (2002-07) they include the following:

- facility to support EPA negotiations (€ 24 million);
- facility to support ACP countries in WTO negotiations (€ 12 m);
- the Pesticides Initiative (€ 29 m); Livestock Disease (€ 72 m); Fisheries Programme (€ 42m);
- Trade.Com (€ 50m) (with its three components being strengthening local trade policy-making, research and training capacities; establishing a network of ACP experts through a so-called "hub and spokes" programme; pilot projects to address urgent institutional and supply-side constraints in technical standards and sanitary and phyto-sanitary requirements).

Recognising that some developing countries are likely to be faced with adjustment difficulties (erosion of trade preferences; reduction in government revenue resulting from lower tariffs, and increased cost of food imports resulting from reductions in export subsidies, for net food importers), the European Community has recently ventured into adjustment mechanisms. They include commodity specific instruments to strengthen competitiveness and cope with erosion of preferences (e.g. on banana, rum, cotton, rice and other agricultural commodities), and more recently the Trade Integration Budget Support Facility (TIBS) under the Eastern and Southern Africa (ESA) RIP in view of revenue losses from East African Community's Customs Union and possible adjustment costs from the EPA (€ 40–50 million).

policies and interventions, adequately bridging the supply and demand sides. These dimensions are crucial for the successful improvement of the quantity and quality of AFT initiatives, as recognised in the international debate on aid effectiveness[18] and in the recommendations of the WTO Aid for Trade Task Force.

4.3.1 Ownership

Lack of *ownership*, that is, the extent to which the beneficiaries of assistance control a programme and its inputs and drive the process, has often been identified as one of the most significant shortcomings of development cooperation.[19]

In terms of ownership, the joint programming dimension of the EDF (see Box 1) is particularly relevant, as the formal CPA procedures explicitly aim at full ownership and local accountability. The EDF programming process is initiated in the respective countries (regions) where the national (regional) authorising officers – NAOs (RAOs) –, along with the European Commission Delegation, prepare and develop a first draft of the Country (Regional) Strategy Paper, "based on the country's own medium-term development objectives and strategies" (Art. 1 Annex IV of the CPA). Moreover, according to the CPA, "the country support strategy shall be an instrument to prioritise activities and to build local ownership of cooperation programmes" (Art. 2).

However, despite some recent improvements, the actual degree of ownership of EDF management is questionable.[20] According to ACP stakeholders, it cannot be taken for granted that EDF assistance is implemented as a real partnership. Evidence shows that "the goal of increasing ownership by partner countries of these processes remains elusive" and that "programming and review processes have not been as 'joint' or 'mutual' as many would like" (Mackie, 2007). The EU donors heavily influence the final allocation of funds and this may bias programming towards the most visible activities, which yield the most tangible results from a donor's point of view, but which could bypass the wishes of ACP communities. Effective and broad participation in defining demands on the ACP side and the depth of needs assessment are questionable, as often only a few officials in the NAO/RAO offices know and participate fully in the programming process (Eurostep, 2006). In the case of aid for trade activities, consultations are mainly conducted with government agencies and regional organisations and very few with private sector and civil society stakeholders (ADE, 2004). The problem is aggravated by the fact that the NAO is generally located at the treasury or at planning ministries, and there is little or no involvement of the trade ministry in the EDF process.

4.3.2 Capacity

Ownership is directly related to the process and governance structure for aid decision-making as well as the *capacity* of recipients to identify priorities for assistance and participate proactively in decision-making processes. Needs identification is a key aspect to ensure ownership. The programming approach adopted by the European Community envisages needs assessment. In the case of the AfT, the analysis of trade performance, production-sector constraints and international commitments, as well as of the role of trade policy within overall development strategies provide an opportunity for more insightful preparation of the European Community aid for trade relative to many other donors. However, a key deficiency emphasised by external evaluations is the lack of rationale underlying many of its aid for trade programmes (ADE, 2004). The analysis by the European Commission of the trade situation in the beneficiary countries is not in-depth; neither do the strategic programmes covering the regions comprehensively analyse the regional trade situation. Identification of ACP partners' needs is done through extensive consultation at the programming stage but apart from a series of meetings, often no real analysis or diagnosis takes place. The European Commission recently adopted a set of guidelines for its AfT, which is expected to significantly improve the programming of these activities.

Building trade capacity relates in general to improving a country's ability to define and pursue its own interests through trade development and trade policy. This implies that sustainable trade capacity is built endogenously with emphasis on human resources and institutional development. A general systemic problem with aid for trade that is broad in scope, as in the case of the European Community, is that by addressing several dimensions and covering so many different types of schemes[21] the extent to which endogenous capacity development actually occurs it is not the focus of the intervention (Rourke, 2006). Short-term assistance needs, including development of 'hard' capabilities (technical skills, trade knowledge, etc.) are generally addressed. But the long-term problem of weak human resources and institutions as well as the lack of 'soft' capabilities (oriented more towards processes, networking and methods) often remains. This may be explained by the fact that as multiple levels of trade negotiations are pressing both donor and recipient policymakers to act (and hard capabilities are easier to identify, develop and evaluate), short-term objectives often dominate the agenda to the detriment of sustainable capacity building (Bilal and Szepesi, 2006). By contrast, changing working methods, relations among key actors and approaches towards trade policy and development are less tangible targets, achievable only with a longer term perspective, which thus tend to be neglected (Solignac Lecomte, 2001).

A particular and significant aspect of the problem of capacity, which also emerges from the experience of EDF programming, is the recipients' weak absorptive capacity. This means in the case of certain ACP countries for instance, that for reasons related to scarce human resources, weak public financial management systems, procedural bottlenecks on the European Community side and perceived lack of ownership amongst some actors, substantial portions of EDF funds are not accessed and remain unspent. For example, significant amounts under the 9th EDF remained uncommitted by 2006.[22]

4.3.3 Monitoring

Monitoring of performance and results is another crucial dimension of development assistance frameworks (Bourguignon and Sundberg, 2007). This, in principle should also include reviewing whether the ownership and capacity of recipients improve over time. Monitoring features prominently in AfT debates and was, for instance, the first aspect to be addressed in the context of the WTO in following up the recommendations of the Aid for Trade Task Force. The WTO Director General Pascal Lamy suggested that the operationalisation of aid for trade should predominantly be evident from monitoring activities, which should take place at several levels.[23]

The European Community aid, and notably the EDF, is subject to regular evaluation and improvement through formal monitoring processes, integrated in the case of the EDF in the rolling programming of the NIPs/RIPs: the annual operational reviews, mid-term reviews and the end-of-term reviews. This is generally considered an advanced system of monitoring and evaluation of aid programmes, especially when compared with other donors' practices, in particular on their AfT activities. Not only is the programming process itself monitored but also the volumes of funds disbursed (Art 1 Annex IV of the CPA). Higher rates of disbursements are perceived as a challenging goal by the EC, with the associated risks though that fast delivery may become an objective in itself (Mackie, 2007).

The focus on regular monitoring and evaluation is the result of the shift from the 'aid entitlements' of the Lomé Conventions to the performance-based partnerships of the CPA. This move was welcomed by many actors (in particular from ACP civil society and private sector) as it provides incentives to ensure credibility (e.g. avoiding hidden agendas and double standards), improve effectiveness (e.g. providing flexible and comprehensive support to good performers) and guarantee simple and transparent decision-making. In practice the mid-term reviews also provide an opportunity to adjust intervention strategies and the corresponding financial resource allocations based on an assessment of both needs and performance.[24]

4.3.4 Lessons on Guiding Principles of AfT

A number of lessons can be drawn from this outline of the ownership, capacity and monitoring issues surrounding management of the European Community AfT, and in particular for the ACP though the EDF mechanism. First, the CPA experience shows that achieving actual ownership of aid programmes by recipient countries is very difficult, even when the parties have explicitly set ownership as an overarching goal of the partnership framework and ownership is pursued through appropriate governance structures and formal procedures. The principle of ownership may remain an ideal if the weak capacity of local stakeholders is not properly addressed.

To ensure real ownership, capacity should be built endogenously with emphasis on long-term human resources and institutional development. In particular, the actual capacity of all relevant stakeholders to participate effectively in aid management processes is very important, and should be part of the focus of the AfT programmes as well. Provision of technical skills (hard capabilities) for short-term needs (e.g. a consultant to assist the NAO in preparing the Country Strategy Paper) may fill some gaps, but contributes very little to the country's actual ability to define and pursue its own development (and trade) interests. The record under the Cotonou Agreement confirms that to improve ownership all involved actors should be directly assisted in learning the exact process and in participating in the cycles of policy and aid programming decisions.

In particular, building trade capacity involves establishing networks that extend beyond government actors. Linkages between the public and the private sector, think-tanks and universities, regional and national institutions are all conductive to improvements in policymaking (OECD, 2001). To make AfT initiatives an owned process, the trade-related capacity of various ministries (including the NAO offices), private-sector organisations and other non-state actors should be strengthened and tailored towards full participation in the programming of aid for trade, which also involves building soft capabilities, such as organisational and networking skills. The capacity-building dimension of any new AfT initiative should therefore include the following:

- participatory approaches to trade policymaking (effective mechanisms for consultation among government, the enterprise sector and civil society);
- support to design a coherent trade strategy that is closely integrated with a country's overall development plan;
- effective mechanisms for intra-governmental policy coordination; a strategy for the enhanced collection, dissemination and analysis of trade-related information;
- trade policy networks supported by indigenous research institutions.

As opposed to 'one-shot' initiatives by means of foreign expertise or thematic training sessions (with special focus on short-term needs linked to ongoing trade negotiations), capacity building in the context of AfT should stress a much longer term commitment by donor and recipient countries with a focus on endogenous change. This would also considerably improve the recipient countries' absorptive capacity.

One aspect of capacity that is especially important to achieve real ownership relates to the ability of both local stakeholders and donor officials to undertake appropriate needs assessment as the starting point for the design of aid programmes. Without a thorough and locally owned[25] diagnostic exercise, even proper governance structures and management skills cannot be translated into effective interventions. As illustrated by the European Community aid for trade practice, more effort is needed from both sides: in the recipient country, for better prioritisation and mainstreaming of its trade and development interests; on the donor side, to move from a reactive approach to partners' demands to a more strategic identification and implementation of trade-related support. To avoid the shortcomings commonly encountered in the ACP-EU case of scant depth of diagnostic analyses preceding trade-related interventions, AfT country and regional strategies should be designed based on a complete trade analysis, in terms of comparative advantages and constraints in the business environment, as well as the endogenously identified trade policies. This would improve AfT programming and increase the contribution of AfT to comprehensive trade supportive reforms.

A further lesson from the cooperation instruments of the Cotonou Agreement in general is that regular monitoring of programmed assistance strengthens effectiveness and, hence, a mechanism for systematic monitoring should be built into the assistance framework and its formal decision-making processes. In the ACP-EU case, improving monitoring systems through the introduction of performance-based mid-term reviews of national and regional EDF programmes has supported improvement of assistance.[26] Analysing such recent experience (mid-term reviews were first undertaken in 2004) might be useful when designing monitoring mechanisms for AfT implementation. Specific lessons may be drawn regarding which actors are best suited to undertake the monitoring, how to ensure follow-up, what exactly should be monitored, what advantages and constraints exist at the national and regional levels for a monitoring mechanism, how monitoring should be designed and funded, and so on.

The ACP-EU experience with performance-based partnerships also shows that any reprogramming should be based on the country's own policy agenda and that needs and performance parameters should be jointly identified by recipients and donors.[27] If the recipient governments do not set targets in a manner coherent with recent trends and policy orientations, the performance assessment will turn into an instrument to enforce conditionality (European Commission, 2002).

An additional lesson is that a pool of funds set aside to reward recipients for performance should be additional to the programmed assistance, to ensure that the mid-term review does not become a zero-sum game in which rewards to 'good performers' are *directly* linked to cuts in allocations to 'non-performers'.

Finally, the EDF mid-term review seems to point to a direct correlation between the choice of sectors and aid instruments, the donor's capacities in a given area and the performance of beneficiary countries. The performance and available expertise at both EC headquarters and within the EC Delegations directly affect an ACP country's performance and should therefore be taken into consideration. Similarly, while designing new AfT packages or evaluating their progress, donors' capacity and specific expertise should be reviewed, with internal reforms put in place when required as part of the aid for trade initiatives themselves.

4.4. INSIGHTS ON THE WAY FORWARD

The experience of the European Community, and in particular in the framework of the ACP-EU partnership, offers useful insights into benefits and shortfalls of aid for trade initiatives in a large scale. Moreover, in a context where the common practice by donors is the country-focus of aid mechanisms, the regional dimension of the European Community aid remains an exception. Indeed, in the framework of North-South relations covering both trade and aid, the European assistance to developing countries, including to the ACP group, entail a regional focus, with substantial resources devoted to regional programmes, including AfT schemes.

It is particularly interesting to analyse the case of European AfT especially given the continued difficulties to substantiate the WTO process on aid for trade. The implementation of the recently adopted *EU Strategy on Aid for Trade* can be regarded as a concrete test on whether the highly publicized new AfT initiatives are going to be 'business as usual' or innovative instruments for more effective trade-related assistance.

A prerequisite for any successful AfT initiative is an evaluation of the strengths and weaknesses of already existing schemes at the national, regional and multilateral levels. Some of the key lessons drawn from aid for trade by the European Community, summarized in this concluding section, could be useful for the assessment of other trade-related schemes and the design of new ones. The insights provided here could serve specifically to identify and address current gaps for regional AfT schemes. Moreover, given that the *EU Strategy on Aid for Trade* is a long term policy framework and the move from pledges to delivery will take place in the years to come, such lessons could contribute to the debate on how to improve effectiveness of the European AfT itself, representing today around half of the world's total trade-related support.

The section concludes with a caveat on the relationship between aid and regionalism. Slow progress in many regional integration processes involving developing countries emphasizes that full ownership of regionalism cannot be given for granted all over the world. Before spending new development cooperation resources, including aid for trade, to support regional integration, a number of questions should be answered on the opportunity of applying the rich nations' model of integration to other countries as well as the relationship between national and regional level aid, including to what extent the latter provides concrete benefits to the poor of individual member states.

4.4.1 Demand, Supply and Management of AfT: Lessons Learnt

Key lessons on how to operationalise the delivery of AfT can be identified, which points to some better practices towards increased aid effectiveness. Such insights derived from the European experience, as summarized below, relate to the principles that should underpin aid for trade, that is, the strengthening of both, the demand and the supply sides of aid during design phases, as well as the improvement of the management and delivery of AfT.

4.4.1.1 Underlying principles *Aid for trade should be based on a partnership approach whereby the parties choose aid and trade as instruments to achieve common objectives, with aid not considered donor compensation (or conditionality) for trade liberalisation by the recipient country.* For trade and aid to work together in delivering on development goals there has to be a shared vision that trade growth and development cooperation, including at regional level, should accompany one another to make a vital contribution to poverty reduction.

Aid for trade should be aligned to the trade priorities endogenously set by developing countries and rooted within their overall national and regional development plans (such as the Poverty Reduction Strategy Papers – PRSPs). Mainstreaming trade into national and regional development strategies is a prerequisite for a holistic, integrated and efficient approach to AfT, but it cannot be forced upon developing countries. Trade should be seen as a major complement to other endogenously designed tools in a poverty-reduction strategy. The broader macroeconomic, productive and social environment has to be taken into account during the programming of AfT interventions. This is essential to increase ownership and credibility of AfT programmes and avoid the perception that the initiative is agency-led rather than a country-led process.

To ensure full ownership by relevant stakeholders in the beneficiary countries, capacity building should include both long-term and short-term interventions with a focus on participation and endogenous change. Full ownership can be achieved only through the long-term commitment by donor and recipient countries to progressively and consistently build the capacity of the various actors involved in AfT to design manage and monitor the AfT process. Aid for trade is likely to address systemic reforms of trade policy and institutions, and trade-related capacity should be strengthened and tailored towards full participation. This will also involve the endogenous development of 'soft' capabilities, such as organisational and networking skills, and not only the short-term horizon-building of 'hard' capabilities (e.g. technical skills and trade knowledge).

4.4.1.2 Improving demand and supply of AfT during design phases *Aid for trade objectives must match existing delivery instruments and actually available resources, and donors should avoid re-labelling existing aid and raising unrealistic expectations.* The extent to which AfT initiatives are perceived as delivering on their promises is key to their success. Too ambitious objectives will generate expectations that cannot be met and create the perception that AfT is ineffective. Thus, a distinction between the initiatives' broader vision and operational objectives, and clarity on actual timeframes for delivery, might be necessary to set up realistic, quantifiable objectives and strengthen the credibility of aid for trade. Classification and reporting methodologies for AfT should be clarified in detail in the early stages of programming assistance, to avoid confusion on the exact scope and amounts and re-labelling of existing aid.

The recognition that trade can make a vital contribution to development should be matched by the broad scope of AfT programmes, both at national and regional levels, and donors' decision to support trade reforms in developing countries with substantial levels of predictable funding. The scope of aid for trade should cover all the factors underlying countries' ability to use trade efficiently as a tool for development (support for trade policy and regulations, trade development, trade-related infrastructure, building productive capacity, trade-related adjustment). AfT volumes should increase together with overall official development assistance (ODA) to avoid shifting in-country priorities and creating competition between different development cooperation areas.

The broader the scope of interventions the more important the proper articulation, definition, timing and reporting of AfT initiatives will be. To be effective, all AfT activities for a single country or region should be designed in a holistic manner and under a coherent framework. Programmes should be broad enough in terms of scope but should retain focus and set clear priorities for each area of intervention. Focus and prioritisation of AfT activities per country and sector might be necessary due to low absorption capacity of recipients and limited financial resources. Decisions on actual targets and timing for delivery might cause resentments if they are not transparent and do not follow clear criteria. This is particularly important in the case of regional aid for trade, suggesting an analysis of the implications of additional aid at regional level for the absorptive capacity of national bureaucracies as well as an assessment of the potential distribution of benefits among members of the same region.

Aid for trade interventions should integrate regional priorities, without losing focus on national-level interventions. Endogenously set AfT strategies should integrate national and regional AfT programmes, since one of the stated objectives of most new AfT initiatives, including at multilateral level, is to assist regional integration. The existence of regional and national AfT programmes will require complementarity and sound coordination between programmes at the two levels. At the same time, this should not mean losing focus at the national level.[28] Strengthening the links and complementarity between the regional and country levels of interventions and identifying national and regional needs is an important step in this direction. The involvement of institutions coordinating regional integration (regional development banks, regional economic cooperation communities and other cross-border secretariats) in preparation for AfT national-level strategies would facilitate the identification of the AfT support needed by each member country to implement the regional integration reforms and commitments. A careful distinction of national and regional needs may require the allocation of adequate specific funds to needs assessment and project design during the AfT programming exercise.

4.4.1.3 Improving the management and delivery of AfT *Given the possibly large amounts to be disbursed and numbers of donors, beneficiaries and projects, aid for trade should be subject to regular evaluation and improvement through formal monitoring and review processes.* A strong monitoring function built into the governance structures for AfT at both national and regional levels would be essential to track progress, ensure the availability and timely channelling of predictable funds and to guarantee efficient use of resources. The needs and performance parameters informing the monitoring and review of AfT programmes should be jointly identified by recipients and donors.

Coordination, complementarity and coherence of aid for trade interventions should be guaranteed through joint programming, and each donor's role needs to be defined on the basis of expertise and comparative advantages. Effective coordination within the donor community, beyond solely the exchange of information has to be ensured. Given the correlation between the choice of sectors and aid instruments, the donor's expertise in a given area and the performance of beneficiary countries, donors' capacity and specific competencies should be reviewed while designing new AfT packages or evaluating their progress. A donor should be able to delegate responsibility for carrying out its aid programme in a particular area to another donor that is better placed to do so. Joint programming and joint financial agreements favouring the harmonisation of procedures are relevant steps in this direction.

Aid for trade resources must be commensurate with objectives and made predictable as well as available in a timely and effective fashion. This implies that modalities and procedures for AfT disbursement must be flexible and user-friendly, requiring improvements on both the donor and recipient side. Timely availability and predictability of funds were fundamental to the success of some of the existing AfT schemes and remain among the most difficult challenges fro any new AfT initiative. As inefficient and ineffective administrative processes stand in the way of timely delivery of support even where the level and scope of aid are appropriate, adequate modalities and procedures could allow for relatively timely and efficient access to AfT resources. The use of budget support and locally owned management procedures (such as partner countries' procurement systems) as aid modalities for AfT would allow for quicker access to resources and facilitate ownership, as well as more predictable but flexible programming.

On the donor side, devolution of aid programming and implementation, from headquarters to the in-country level, is crucial. This includes increasing regular communication between donor headquarters, donor field missions and operational colleagues and systematic involvement of field staff in decision-making. In addition, reporting and auditing requirements should be simplified and procedures should allow for recruitment of local staff. On the recipient side, governments should strengthen local institutions and aid-management frameworks for transparency and financial accountability, and take the lead in enhancing communication with donors, implementing agencies and stakeholders. Moreover, beneficiaries should clearly define responsibilities, decision-making authority, and task divisions among the actors involved in managing AfT at national and regional levels.

In the quest for balanced spending between regional and national level aid for trade and more efficient aid modalities, the case of the EU Strategy on Aid for Trade is particularly important to test possible improvements in terms of division of labour, harmonization and alignment among donors. The EU Strategy builds upon two important pillars of European development policy, in the process of being strengthened: the regional level support and the adherence to international aid effectiveness principles. In a situation where the EU is made up of a multitude of donors and development agencies and those collectively constitutes already the world's major provider of regional AFT, establishing a common EU approach to AFT is an important opportunity. In the case of the six ACP regions that should be among the first beneficiaries of the EU Strategy on Aid for Trade, for instance, few EU member states are active at regional level, whereas the European Community is a major provider of regional programmes. This offers an opportunity to ensure an effective division of labour, especially if EU member states can overcome their reluctance to rely on other EU donors programmes in regions where they are not present.

It has to be seen whether EU member states and the European Commission can agree to pool their resources and thus exploit concretely the possibilities for better division of labour in terms of using co-financing, voluntary contributions and joint implementation arrangements, especially at the regional level. Division of labour in particular should be based on a careful assessment of different comparative advantages in different assistance areas by different donors. A donor should be able to delegate responsibility for carrying out its aid programme in a particular area to another donor that is better placed to do the job. For instance, the European Commission seems better equipped to manage programmes which contain large investment components than programmes that depend on processes (i.e. trade governance, capacity building, etc.), so the latter could be left to other EU donors with better performance in these areas. Implementation of the EU Strategy on Aid for Trade should include, as a prerequisite, diagnostic exercises to determine which donor should focus on which part of the AfT package in each ACP region and country. As in the ongoing discussions between EU member states and the European Commission, this also entails a joint financial agreement favouring the harmonisation of procedures around a 'country system'. In terms of regional AfT, harmonization and alignment could be strengthened by using the concept of 'lead donors' to ensure that the EU speaks with one voice when it comes to regional aid for trade and by supporting regionally owned AfT mechanisms as key delivery modalities[29].

4.4.2 Promoting Regionalism: Conditionality or Endogenous Strategy?

The lessons identified above could be useful to improve the effectiveness of aid for trade, starting with the implementation of the EU Strategy on Aid for Trade itself. However, by emphasizing the importance of full ownership of policies and programmes adopted by recipients for the effectiveness of accompanying aid, they also point to a caveat of particular importance in the case of regional aid for trade. Almost by definition regional AfT is about strengthening regionalism, through regional coordination of policies, strengthening regional organizations, building regional infrastructure and markets. But lack of commitment by involved actors and slow progress in many regional integration processes involving developing countries emphasizes that full ownership of regionalism cannot be taken for granted. Many, with various arguments, have warned about using the success of North-North integration, starting with the EU case, to push countries in the South to take the similar policy direction (Bilal, 2004). In situations where regional integration processes lack full ownership by individual member states, donors' willingness and support to foster regionalism risk to be perceived as conditionality to obtain larger volumes of aid.

Before spending new development cooperation resources, including AfT, to support regional integration, a number of questions should be answered on the opportunity of applying the rich nations' model of integration to other countries as well as the relationship between national and regional level aid, including to what extent the latter provides concrete benefits to the poor of individual member states. This is not to deny the importance of filling the gaps in situations when more regional AfT is needed, but to warn on the risks of considering assistance at regional level a priority and necessarily beneficial for all countries at any moment in time. The case of ACP-EU is particularly relevant. On the one hand, drawing on its own experiences, the EU pushes regionalism in the ACP, with trade and support to regional integration as longstanding development priorities for the EU.[30] On the other, there is evidence that this enthusiasm is not shared by most ACP recipients[31].

Official statements indicate that nearly all developing countries have decided long time ago to integrate with neighbours. Recent policy directions confirmed the importance to undertake regional integration. In the case of Africa for instance, the four pillars of the Joint EU-Africa Strategy are 'trade and regional integration', 'peace and security', 'governance and human rights', 'key development issues'.[32] However, policy statements often differ from reality. An indication of the relative importance of trade compared to *non* trade-related interventions emerges from the current choices made by African governments for the 2007 programming of the EDF at national level. Very few AfT programmes can be detected as part of the focal sectors identified in the 31 strategy papers available for African ACP countries (ECDPM and ODI, 2008). At national level, where most aid for trade needs will be addressed, governments choose to allocate funds outside the AfT categories. Similarly, despite the increasing interest in regional support from all corners[33] and high-level policy statements, by early 2008 virtually no ACP region had adopted a regional AfT strategy.

This, coupled with the unsolved constraints to production and trade capacity at national level, the problematic coordination of policies and aid absorption capacity at regional level, as well as the institutional weaknesses of regional organizations and actors, suggest an honest assessment of the degree of ownership for regional integration processes and the adequate prioritization of national vis-à-vis regional AfT programmes. A lot remains to be done on the side of developing countries to avoid falling into the trap of conditionality and to truly mainstream regional integration and trade into national and regional development plans. The ball is squarely in their camp to identify a realistic set of priorities, based on political choices and sound analysis of existing efforts. Thematic and geographic priorities should guide the scope and specific sectoral allocations of aid for trade, not donors' willingness.

To avoid failure of regional AfT due to non-commitment to regional reforms despite availability of support from donors, developing countries should 'drive' the process ensuring they have an endogenous plan as part of which a donor's strategy comes in with operational resources. Concretely, following national/regional aid for trade strategies, there is a need for:

- stocktaking of existing AfT at regional and national levels;
- identification of gaps (both in overall financing needs and sectoral aid allocations) and delivery instruments;
- proposals for where/how interested donors should contribute with AfT resources.

If such steps are not taken in this order, there is a risk of perpetuating the lack of full ownership. When deciding on how to deliver on their aid for trade pledges, donors could help by devoting part of their AfT funds to strengthening ownership, especially in the early stages of the process. Resources could support both capacity building to design AfT demands and the domestic processes for mainstreaming owned aid for trade strategies and programmes in national and regional development strategies.

NOTES

1 See for instance Bilal and Szepesi (2006) and Stiglitz and Charlton (2006).

2 See for instance OECD (2006 and 2007).

3 This Chapter is a revised version of the Marti and Rampa (2007).

4 See Council of the European Union, "EU Strategy on Aid for Trade: Enhancing EU support for trade-related needs in developing countries", 15 October 2007, http://register.consilium.europa.eu/pdf/en/07/st14/st14470.en07.pdf

5 The Paris Declaration, endorsed on 2 March 2005, is an international agreement to which over one hundred ministers, heads of agencies and other senior officials adhered and committed their countries and organisations to continue to increase efforts in harmonisation, alignment and managing aid for results with a set of monitorable actions and indicators.

6 In October 2006, the EU formally decided to address EPA-related adjustment needs within the broader framework of aid-for-trade; See "Conclusions on Aid for Trade", 12 October 2006, Council of the European Union http://register.consilium.europa.eu/pdf/en/06/st13/st13882.en06.pdf

7 See European Commission (2006).

8 See ADE (2004).

9 See http://tcbdb.wto.org

10 See the General Affairs and Economic Relations Council of the European Union (GAERC) 'Conclusions on Aid for trade' of 12 October 2006 and the GAERC 'Conclusions on EPAs' of 26-27 May 2008.

11 Reporting and classification uncertainties are common to several donors (see OECD 2007 for details). Some AfT activities, especially to support the private sector, could be classified under several of the categories of the WTO Aid for Trade Task Force, depending on how exactly these are defined.

12 Re-labelling all growth-promoting development assistance as AfT would reduce the impact of the AfT initiative and increase scepticism about its potential to effectively enhance developing countries' ability to better benefit from trade liberalisation (OECD 2007).

[13] According to the EU Strategy for Africa, "The 10th EDF will support the implementation of EPA[s]. The main focus of regional programmes will be regional economic integration, EPAs and trade" (see http://www.europe-cares.org/africa/partnership_next_en.html).

[14] See ILEAP (2007) for an overview of current regional and national aid for trade programmes supported by the European Community.

[15] It should not be overlooked, however, that there may be a trade-off between recognising the importance of AfT and increasing resources for it and reducing levels of support to other development objectives (like health, education, etc.). This risk that funds may be diverted from other development programmes to AFT has been expressed by ACP governments (ACP 2006) as well as civil society (FTAO 2007).

[16] This risk is recognized by the European Commission: "TRA in the aggregated data for the ACP are dominated by regional and all-ACP programmes in part designed to prepare for the EPAs. There are only relatively few ACP countries that have a trade programme, even though this number is increasing reflecting increased attention for trade and development over the past few years. In contrast to the ACP countries, the share of TRA within country programmes is higher in Asia and in Latin America." (European Commission 2005).

[17] See also for instance ADE (2004), European Commission (1999 and 2002), Eurostep (2006), Frederiksen and Baser (2004), and Mackie (2007).

[18] An overview of recent evidence suggests that the two central features of effective aid architecture are country ownership of the development strategy and allocation of aid on the basis of performance (Bourguignon and Sundberg 2007).

[19] "For instance, in the case of multi-donor initiatives, such as the Integrated Framework, partner countries often failed to demonstrate strong leadership and political will for the reforms needed to underpin an effective trade development strategy, and the trade agenda seemed to have been driven mainly by the donors" (OECD 2006).

[20] For a thorough discussion see Mackie (2007).

[21] Amongst these are fiscal revenue loss, implementation of international commitments, preparation for negotiations and trade-policy formulation, private sector development support, trade infrastructure and preference loss.

[22] In April 2006 ACP finance ministers "expressed concern that there is still [a] substantial uncommitted amount under the 9th EDF", and urged "the ACP States and Regional Organisations to urgently consider the reallocation of these resources to existing and new instruments addressing the challenges of market liberalization, preference erosion and the process of integration into the world economy so that these funds are not lost under the sunset clause in December 2007" (ACP 2006).

[23] "Monitoring to assess global aid for trade flows (using the OECD-DAC Creditor Reporting System database to increase transparency on pledges and disbursements); monitoring the evaluation reports by development agencies; and monitoring and evaluation of in-country aid for trade." (Director General's Report on Aid for Trade to the General Council, 14-15 December, 2006JOB(06)262, WT//1, http://docsonline.wto.org).

[24] Needs parameters include per capita income; population size; social indicators; indebtedness; export earning losses; and dependence on export earnings, in particular from the sectors of agriculture and mining. Performance parameters include progress in implementing institutional reforms; country performance in the use of resources; effective implementation of current operations; poverty alleviation or reduction; sustainable development measures; and macroeconomic and sectoral policy performance (Frederiksen 2003).

[25] Most of the recent evaluations of AfT programmes highlight that "bilateral donors have not always assessed trade-related needs in consultation with all relevant stakeholders, e.g. the private sector, while non-governmental stakeholders were not systematically consulted during the design and implementation of bilateral or multilateral programmes" (OECD 2006).

[26] For a thorough discussion see Frederiksen (2003).

[27] In the context of AfT of course 'performance-based' should not mean that aid is disbursed on the basis of performance in terms of trade liberalisation by the recipient country. This 'stick and carrot' approach would undermine the credibility and effectiveness of AfT initiatives.

[28] This risk is recognised by the European Commission: '[Trade-related assistance] TRA in the aggregated data for the ACP are dominated by regional and all-ACP programmes in part designed to prepare for the EPAs. There are only relatively few ACP countries that have a trade programme, even though this number is increasing reflecting increased attention for trade and development over the past few years. In contrast to the ACP countries, the share of TRA within country programmes is higher in Asia and in Latin America.' European Commission (2005).

[29] The European Commission has already started providing aligned support to several ACP regions. The (9th EDF) aid has been aligned with the systems, procedures and modalities of the Regional Economic Community (RECs) through the use of so-called 'Contribution Agreements'. Only RECs or international organisations which have in place procedures and governance systems which reflect international standards are eligible. Eligibility has been verified through institutional audits focused on procurement, internal/external audit, financial – and non-financial reporting and internal governance structures.

[30] In recent years, "for the ACP, at least a quarter of TRA have been for regional integration, specifically, and many TRA programmes are carried out at all-ACP level" (OECD and WTO 2007).

[31] For a discussion about stakeholders' perceptions on regional integration and the relationship between national and regional level aid in the Pacific, see Rampa (2007).

[32] The Joint Strategy, which provides an overarching long-term policy framework for Africa-EU relations, will be implemented through successive Action Plans which will build on the operational part of this Joint Strategy and cover proposed priority actions for three years. Under the First Action Plan (2008-2010) one of the priorities is the Africa-EU Partnership on Trade, Regional Integration and Infrastructure.

[33] Regional approaches to aid for trade were also identified as crucial in July 2006 by the WTO Task Force, which urged: "donors and agencies, together with regional banks and organizations, to step up their efforts to identify regional, sub-regional and cross-border needs, including those related to regional integration"; "to improve the matching between recipients' demands and donors' responses also on regional, sub-regional and cross-border issues".

REFERENCES

ACP (2006). Declaration of the 3rd meeting of the ACP Ministers of Finance and Economic Affairs, 28 April 2006, Brussels. http://www.acpsec.org

ADE (2004). Evaluation of Trade Related Assistance by the EC in Third Countries Vol. 1 & 2, Louvain-la-Neuve. http://ec.europa.eu/europeaid/evaluation/eval_reports/reports_by_year.htm#2004

Bilal, S. (2004). Can the EU Be a Model and a Driving Force for Regional Integration in Developing Countries?, Paper presented at the Second Annual Conference of the Euro-Latin Study Network on Integration and Trade (ELSNIT), Session III – Institutional Dimensions of Integration and Trade, 29–30 October 2004, Florence. http://www.iadb.org/europe/index.php?file=news_and_events/elsnit_conf_2004/Bilal_ELSNIT_2004.pdf

Bilal, S. and S. Szepesi (2006). Trade Capacity Building as a Facilitator of Global Public Goods. In: Secretariat of the International Task Force on Global Public Goods (Ed.), *Meeting Global Challenges: International Cooperation in the National Interest*, Stockholm, Sweden: International Trade. http://www.gpgtaskforce.org

Bourguignon F. and M. Sundberg (2007). Aid Effectiveness – Opening the Black Box, World Bank paper presented at the American Economics Association annual meeting, January 5–7 2007, Washington. http://www.gsdrc.org/go/display&type=Document&id=2614

ECDPM and ODI (2008). The new EPAs: comparative analysis of their content and the challenges for 2008, ECDPM Policy Management Report 14, Maastricht. http://www.ecdpm.org/pmr14

European Commission (1999). Development and Humanitarian Assistance of the European Union. An Evaluation of the Instruments and Programmes managed by the European Commission", Final Synthesis Report, Brussels.

European Commission (2002). Progress Report on the Implementation of the Common Framework for Country Strategy Papers, Commission Staff Working Paper SEC(2002)1279, Brussels.

European Commission (2005). EC Trade-Related Assistance (TRA) – some key facts and figures, Brussels: European Commission.

European Commission (2006). Making trade work for development, Brussels: European Commission. http://trade-info.cec.eu.int/doclib/html/118805.htm

Eurostep (2006). An assessment of the Programming of EC aid to ACP countries under the 10th EDF, Brussels: Eurostep. http://www.eurostep.org/wcm/index.php?option=com_docman&task=doc_download&gid=98&Itemid=41

Frederiksen, J. (2003). Mid-Term Reviews: Performance-based partnerships in ACP-EU cooperation, *ECDPM InBrief*, No. 5. http://www.ecdpm.org/inbrief5

Frederiksen, J. and H. Baser (2004). Better aid delivery or deconcentration of bureaucracy? a snapshot of the EC's devolution process, *ECDPM InBrief*, No. 10. http://www.ecdpm.org/inbrief10

FTAO (2007). Contribution from the Fair Trade movement to the EU debate on 'AfT', Fair Trade Advocacy Office, April 2007. http://www.fairtrade-advocacy.org

General Affairs and Economic Relations Council of the European Union (GAERC) (2006). *Conclusions on Aid for trade*, Brussels: European Union (12 October 2006).

General Affairs and Economic Relations Council of the European Union (GAERC) (2008). *Conclusions on EPAs*, Brussels: European Union (26–27 May 2008).

ILEAP (2007). Regional Aid for Trade, *ILEAP Negotiating Advisory Brief*, No. 12. http://www.ileap-jeicp.org

Mackie, J. (2007). EDF Management and Performance. In: G. Laporte (Ed.), *The Cotonou Partnership Agreement: What role in a changing world? Reflections on the future of ACP-EU relations*, ECDPM Policy Management Report No. 13. http://www.ecdpm.org/pmr13

Marti, D.F. and F. Rampa (2007). Aid for Trade. Twenty Lessons from Existing Aid Schemes, ECDPM Discussion Paper No. 80.

OECD (2001). *DAC Guidelines on Strengthening Trade Capacity for Development*, Paris: OECD. http://www.oecd.org/document/31/0,2340,en_2649_34665_1886111_1_1_1_1,00.html

OECD (2006). Trade-Related Technical Assistance and Capacity Building: What Do Recent Thematic Evaluations Tell Us?, COM/DCD/DEV(2006)1, Paris. http://www.oecd.org/dataoecd/31/45/37624158.pdf

OECD (2007). Aid for Trade: Making it Effective, Development Co-operation Report, Vol. 8 (1), Paris. https://www.oecd.org/dataoecd/23/15/37438309.pdf

OECD and WTO (2007). Aid for Trade at a Glance: Country and Agency Chapters. Paris: OECD http://www.oecd.org/document/46/0,3343,fr_2649_34665_39619566_1_1_1_1,00.html

Oxfam (2005). Scaling Up Aid for Trade: How to Support Poor Countries to Trade their Way Out of Poverty, Oxfam Briefing Note, 15 November.

Rampa, F. (2007). Implementation of Article 37 (4) of the Cotonou Agreement – Provision of technical support to assist the Pacific ACP Region in the Review of EPA Negotiations, Maastricht: The European Centre for Development Policy Management. http://www.ecdpm.org

Rourke, P. (2006). Strategic positioning and trade-related capacity development: The case of CTPL and Russia, ECDPM Discussion Paper 57O, Maastricht. http://www.ecdpm.org/dp57o

Solignac Lecomte, H.-B. (2001). Building capacity to trade: a road map for development partners – insights from Africa and the Caribbean, ECDPM Discussion Paper 33, Maastricht. http://www.ecdpm.org/dp33

Stiglitz, J. and A. Charlton (2006). *Aid for Trade, a report for the Commonwealth Secretariat*. London: Commonwealth Secretariat.

CHAPTER 5

AID FOR TRADE FOR SUSTAINABLE DEVELOPMENT

5.1. INTRODUCTION

This chapter presents a constructively provocative set of ideas on how the concept of 'Aid for Trade' (AfT) can be understood within the larger context of sustainable development. It is suggested that not only can AfT contribute to the pursuit of sustainable development but that sustainable development, which is now the stated goal of international environmental and World Trade Organisation (WTO) policies, should be amongst the meta-goals around which AfT initiatives and investments are structured.

Instead of repeating the minutia of specific environmental concerns, the first section of the chapter sets for itself the goal of laying out the broad conceptual argument for why AfT has to have a broader goal than just the facilitation of increased trade or enhanced trade liberalisation. Although sustainable development is clearly a central policy goal for all of the activities of the United Nations Environment Programme (UNEP), the chapter avoids making the parochial argument for sustainable development. Instead, it highlights the fact that sustainable development is now the shared policy goal for *all* actors involved in AfT discussions. Hence, sustainable development is not just an environmental concern, but equally a development priority and now the stated policy goal of trade policy (Najam et al., 2007).

This is followed by the distillation of three key lessons that this chapter proposes should guide the actual planning of AfT initiatives. It is proposed that any future discussions and any framework for AfT that emerge from those discussions should build upon these three lessons. In particular: (a) general lessons from international assistance experience should be examined for additionality, predictability and conditionality, and to create processes of access that are transparent, relevant and needs-based; (b) a key priority for AfT initiatives should be the cultivation of domestic capacities for integrated analysis and assessment; and (c) AfT investments should particularly focus on capacity enhancements for policy processes and implementation, especially in the areas of international policymaking capacities, domestic implementation capacities, and stakeholder participation capacities.

The lead author for this chapter is Dr. Adil Najam. He is the Frederick S. Pardee Chair for Global Public Policy and a Professor of International Relations and of Geography and the Environment at Boston University, Boston. The views, analysis and interpretations expressed in this paper are those of the author and do not necessarily represent the official position or opinion of UNEP or its member countries, nor does citing of trade names or commercial processes constitute endorsement.

P. De Lombaerde, L. Puri (eds.), *Aid for Trade: Global and Regional Perspectives*,
United Nations University Series on Regionalism 2,
DOI 10.1007/978-1-4020-9455-2_6,

The chapter concludes with thoughts on the institutional strategies and collaborations between the United Nations Environment Programme (UNEP) and other agencies that could help realize the strategic approach laid out in this chapter.

5.2. THE LOGIC OF AID FOR TRADE FOR SUSTAINABLE DEVELOPMENT

The genesis of the notion of 'Aid for Trade' – or AfT – lies largely in the realisation that materializing the potential benefits of trade is not simply a matter of liberalizing regulations (Halle, 2005; Rodrick, 2001) It also requires an enabling policy environment and building of the full spectrum of critical capacities – institutional, informational, infrastructural and implementational (WTO and OECD, 2007).

This conceptual realisation began to set in soon after the 2001 Doha Ministerial Conference of the WTO, and by December 2005 had become a key item and recommendation in the WTO Hong Kong Ministerial Declaration (Hoekman, 2002; Heydon, 2006). It is not surprising that the notion is now widely embraced by both industrialised and developing countries. Indeed, the idea was fully accepted not only for conceptual reasons but also for its political rationality. By the end of 2005, the Doha Development Agenda negotiation process initiated in Qatar in 2001 had already faltered, as the early fascination with trade liberalisation for its own sake began to fade, particularly amongst many developing countries. Some gesture of North-South cooperation to ensure that the benefits of trade extended to everyone and across generations was dearly needed (Heydon, 2006).

It also came along at the exact time to meet environmental, or more precisely, sustainable development concerns. The sustainable development justification for AfT arises from the synergy between the trade logic and the environmental logic for AfT. On the one hand, the logic for international trade policy flows from the recognition that increased investment is needed in developing countries to capture net development gains from greater global trade and liberalised world markets. Without the needed capacities and conducive internal policy conditions, developing countries and economies in transition are not able to harvest the benefits of increased market access and may even find themselves facing new trade disadvantages with liberalised trade competition. On the other hand, the environmental logic for AfT builds on an ongoing trade and environment debate that is now nearly 15 years old. This vigorous debate has arrived at a conclusion that trade and environment policies not only needed to coexist, they would actually need to cooperate for the achievement of the broader goal of sustainable development. Therefore, sustainable development is not only the meta-goal of environmental policy but also of trade policy (as outlined in the WTO Preamble and more recently in the Doha Declaration) (Halle and Meléndez-Ortiz, 2007).

The sustainable development logic for AfT draws together both trade logic and environmental logic, and then broadens that within the context of development and poverty alleviation. It builds on the conceptual realisation that compartmentalizing trade and environment policy and handling them separately is neither feasible nor desirable. Firstly, much of environmental policy has important trade implications, and a significant part of global environmental policy is trade policy in itself (Najam et al., 2007). For example, trade is a key component of the management of endangered species, and CITES is as much of a trade agreement as it is an environmental treaty. In the case of biodiversity, the Cartegena Protocol on biosafety is relevant but there are others. The links between trade and environment are deep and mutually influential. Biodiversity and biodiversity-related products are critical trade commodities while trade is a critical instrument in the management of global biodiversity. Secondly, it is evident that within an entire array of issues such as global waste trade, resource use efficiency, ecosystem management, sustainable production and consumption, organic agriculture, fisheries, etc., the concerns about the global environment are congruous with the concerns about global trade in numerous ways. Trade can both exacerbate environmental challenges and address them, and trade and environmental policies must both aim to create the conditions where the goals of trade and of the environment are met simultaneously. However, the case for sustainable development broadens the challenge because it demands that both trade and environment policies are sensitive to issues of development particularly for the poorest and most vulnerable populations. These are the populations that are most likely to bear the costs of both policy domains and, therefore, their concerns should be heard (Najam et al., 2007).

The sustainable development logic for AfT recognizes that AfT can be an effective tool to induce greater policy coherence domestically as well as internationally, and also encourage policy synergies, especially those articulated in the Doha Mandate, for the achievement of sustainable development (Genee, 2007). The promise of sustainable development has always had the premise of bringing together developmental and environmental action and creating a combined rationality for the two (Nijnkeu, 2007). AfT provides a potential instrument for doing so in the case of trade and environment (WTO and OECD, 2007)[1].

A point of emphasis is that neither trade liberalisation nor increased trade volumes is a policy *goal* in itself. These policy goals become worthy in only so much as they lead, or promise to lead, to improvements in the quality of human well-being and the quality of the environment in which human well-being can be sustained, or sustainable development. In this sense trade has to be viewed as a policy *instrument* much more than a policy *goal*. The purpose of policy is not to increase trade, *per se*. The purpose of policy is to enhance trade so that something else – in this case, sustainable development – can be achieved. Aid, of course, is also a policy *instrument*. Even though 'more aid' is generally considered a good thing, just as 'more trade' is a generally perceived as a good thing, the fact of the matter is that ultimately the quantum of aid is an incomplete measure of the efficacy of aid. The effectiveness of aid can only be measured, in the ultimate sense, by the degree of change from the status quo that has been brought about as a result of that aid (Najam and Robins, 2001).

The point to be made here is that in putting 'aid' and 'trade' together, AfT combines the two most potent 'means' or 'instruments' of policy. But for what purpose? Sustainable development, as the professed goal of the global developmental and environmental communities, is the 'end' or 'goal' that AfT seeks to achieve. In very practical ways, it is not enough to say, "Aid *for* Trade." That only begs the question "Trade for what?" Sustainable development provides a possible answer to that question – "Aid *for* Trade *for* Sustainable Development" (AfTfSD). This paper suggests that aid should be used to instil the conditions for trade that will lead not just to more development, but to more sustainable development including a greater focus on environmental concerns and poverty alleviation. After all, it is the latter (sustainable development) that is now the stated goal of the global trading system, as articulated in the WTO agreement and elaboration in the Doha Declaration.[2]

5.3. LESSONS FOR DESIGNING AFT INITIATIVES

The practical implication of all of the above is that the ultimate purpose of AfT cannot simply be to invest in aid that will create the conditions for more trade and for more trade liberalisation. The ultimate purpose should be to invest in aid that will create the conditions for trade that will lead to more sustainable development; with a special focus on environmental improvement and poverty alleviation. This implies that for AfT to yield concrete gains in sustainable development, it not only has to build on what is known about international assistance (aid), international trade, and environmental stewardship in a sustainable development context, but also combine it (Najam et al., 2007).

At least three key priorities can be determined for those who are to design AfT initiatives in the context of sustainable development. These are necessarily broad-based and apply to all assistance and not just AfT. Without attention to these, agendas will tend to falter. Attention is also needed in the specifics of each context and the cases to which these ideas are applied. These are amongst the key lessons that are derived from the common area of understanding of international aid, international trade policy, and international environmental policy.

5.3.1 Doing Aid Right

The first set of lessons that both trade and environmental policymakers need to highlight relates to what is known about 'aid' (Riddell, 2007; Collier, 2007). There is a long track record of trying to achieve policy goals through international assistance. Much has been achieved when using this instrument; but the literature on the subject also suggests that it was probably not as much as one had hoped for (Banerjee, 2007). The literature also points out that the formula for effective aid is uncertain, although it is quite evident that there have been a few lessons learnt on the way about things that should and should not be done in designing any aid program. These lessons are not unique or specific to AfT and should ideally be applied to all international assistance. Here are three lessons in particular that any designer of AfT would be well advised to keep in mind.

5.3.1.1 Additionality and predictability Developing countries have long worried that promises of assistance in one area will be fulfilled but only by cutting off assistance in other areas. This is an issue of *additionality*. Because aid is never an obligation on the donor and because generosity can tend to be fungible, it is all too easy for a donor to move assistance from one area into another. Moreover, the case for additionality also comes from the fact that trade rules can erode existing preferences enjoyed by developing countries and that shock can be significant to fragile economies. AfT can then become a mechanism to ease the transition and the shock from change (Hoekman and Prowse, 2005; Wijayasiri and Samad, 2007).

The literature on the subject suggests that from the recipient's perspective such a mindset only exacerbates policy confusions, confounds efforts that may already be in place, and shifts aid towards a donor-driven agenda rather than a need-driven set of priorities (Tarp, 2000). All estimates suggest that considerable capacity enhancements are needed in AfT – institutional, infrastructural, informational and implementational – and they must come from additional sources and not compete with existing assistance expenditures in the environmental, trade or other related arenas (UNCTAD, 2007). This, of course, is intimately connected to the issue of *predictability*. Aid fashions can be extremely fickle and issues that are supposedly 'priorities' today can fade tomorrow simply because the aid community and donors have found a new favourite cause (Collier, 2007). Meanwhile, recipient countries might already have set actions and processes in motion based on the assumption of continuing support. Many investments needed in the realm of AfT are institutional and infrastructural and require long-term commitments. These commitments are more likely to be made by developing countries if they know that the commitments for the required resources are going to be sustained.

5.3.1.2 Conditionality Although also related to the issue of additionality, the lessons about conditionality need to be addressed separately because they are of special significance in the trade arena (Dijkstra, 2002). Unlike many other areas of international policy, and certainly unlike much of international development policy, trade policy tends to be "rule-based" (Cline, 2004). It is regulatory in ethos and tends to be based on clear obligations as well as clear penalties for not meeting those obligations (i.e. dispute mechanisms). The tendency of the trade policy world, therefore, is always to push for "obligation or rule-based" systems, since these have generally served the trade regime well. The problem is that what is termed as "obligation" in a trade system can look very much like "conditionality" in a development assistance system. Conditionality has not worked well in the development arena. There is a clear conceptual reason for this disconnect. "Obligations" or rules work when they are mutually created and implemented by all parties. "Conditionality," however, tends not to work because recipient countries view it as an imposition and because it is not implemented by all parties. Indeed, aid relations cannot be "equal" in meaningful ways and therefore cannot be implemented equally. The temptation to cloak conditionality as "obligation" or "rules" must therefore be resisted in the architecture of any AfT obligation. Clear goals, targets, timelines, and monitoring systems can certainly be set up for both recipients and donors, but the suggestion of political or policy conditionality should be avoided (Bayne, 2004).

5.3.1.3 Access based on transparency, relevance and need The third related issue is about access to aid. Countries do realize that aid is never a right for recipient countries and that for donor countries, aid is not simply a development instrument; in some cases it is also one of the foreign policy instrument (Cline 2004). However, in designing an AfT regime, in terms of who can access AfT resources, and how and for what reasons, clarity on access criteria can be critical to the efficacy of such initiatives. At the most basic level, it is a question of *designing a transparent system* of access for AfT resources so that recipient countries have a clear and transparent understanding of what resources are available, for what purposes, and how they may be accessed.

In addition to transparency, the system of access must also demonstrate relevance. Access to AfT resources should be based upon a demonstrated relevance to trade goals, especially to trade goals within a sustainable development context. Finally, and most importantly, *access to AfT resources should be needs-based*. Different countries and the different sectors within these countries will have differing needs for AfT support; the level and nature of this need should be critical criteria for access to such resources. In practice, the least developed countries are likely to show the greatest need for financial and technical assistance. However, they are not the only countries that will have such needs, and the quantum and nature of needs will then again vary within each country and between the countries. Any criteria developed for managing AfT resources should account for this diversity in needs, and match their relevance carefully and transparently to the larger trade and sustainable development goals that are being furthered through any particular AfT investment (Page, 2007).

5.3.2 Building Integrated Analytical and Assessment Capacities

A lack of domestic analytical capacities for trade policy assessment, and for the assessment of its links to environmental and social considerations, not only leads to less than desirable policy outcomes, it also contributes to a sense of distrust and unease with international agreements in general. A sound national analytical infrastructure will, therefore, contribute not only to improved internal and international policy, but also contribute to a deeper buy-in and a sense of ownership for policy. The latter can directly trigger more vigorous and effective implementation.

The need to invest AfT resources into building domestic analytical and assessment capacities in developing countries is all the more important in the case of sustainable development – which is, after all, a cross-sectoral priority by definition. Even those developing countries that may have reasonable domestic analytical capacities in some or all of the component areas of sustainable development often lack the capacity to conduct cross-sectoral assessments of trade policy impact within the sustainable development framework. A critical need, therefore, is for integrated assessment of impact, and for the development of domestic capacities to conduct such assessment and to feed it into national and international policy (Abaza, 2007).

Sustainable development would require us to move beyond traditional sectoral approaches to development policy, and towards integrated assessments which bring together the economic, social and environmental dimensions and analyse the net development impact of trade policy across, and on the intersection of, these dimensions. For example, the development of policies that integrate environmental and social considerations into trade policy could enhance national capacities to access markets (e.g., through trade of environmentally friendly products such as organic agriculture, which are beneficial for the environment, effective in poverty alleviation and good for trade) (Nadal, 2007). There is, for example, now a growing focus on how global trade and global climate change might affect each other. Exploring and exposing these links will require greater emphasis on integrated assessments (Rotmans and van Asselt, 2004). Such analysis can also encourage policymakers to move towards investments in environmental technologies, by eliminating subsidies to polluting technologies or products and channelling these gains towards more sustainable investments.

The examples of policy effect can be numerous, but the core rationale is that AfT initiatives should prioritize investments that allow countries to develop the analytical and assessment capacities needed for the development of integrated policies. These policies are based on a full understanding of the economic, environmental and social linkages and interactions that lie at the core of sustainable development. Such assessments and analysis requires wide-ranging consultation amongst policy and practice stakeholders and those who have the requisite knowledge of the different dimensions. AfT investments are particularly well-suited to integrated assessment because the external resources are needed for national systems that are traditionally structured around sectors (Abaza, 2007).

There is much discussion in the literature on trade policy, and also on development policy, of the importance of "mainstreaming" trade policy into development policy, or vice versa. This is, indeed, a desirable goal. However, there is also a need to mainstream environmental policy into trade policy, and vice versa. Indeed, the mainstreaming argument is valid for multiple sectors in multiple directions. Integrated assessments provide a practical and demonstrated way to integrate all the "mainstreams" into a combined framework that builds upon each domain. A particular manifestation of the type of integrated assessments that is being highlighting here would be a set of trade policy impact assessments as a tool for assisting countries in their negotiating strategies as well as a tool for implementation design (Najam, 2004).

Although the critical importance of policy coherence, itself a stated goal of the global trade policy, will be discussed in the next section, an ancillary benefit of AfT investments in developing analytical and assessment capacities in developing countries, especially for integrated assessment, is that it will cultivate a broad human resource base within these countries and will encourage greater interaction and coherence between the relevant policy domains. Indeed, it could be argued that by its very structure integrated assessment is itself a motivator of enhanced policy coherence amongst different sectors (Wing Yee Lee, 2007).

5.3.3 Investing in Policy Processes and Implementation

As discussed above, the logic of AfT flows directly from the realisation that key institutional capacities and epistemic communities need to exist within a society before the benefits of trade liberalisation can begin to accrue. A key priority for AfT investments, therefore, has to be the development of such capacities (Najam, 2004). The need for analytical and assessment capacities that has already been discussed above is a key input into more direct policy capacities which will be discussed here. In the context of sustainable development, policy capacities need to be developed on three critical levels.

5.3.3.1 International policymaking capacities The most obvious level of capacity-building, a level which already attracts some, but not enough, resources, is that of international policymaking. Most attention is usually paid to capacity-building related to the substance and process of international negotiation (Najam, 2004). These are key aspects and will require more attention in the future. However, there is a need to broaden this understanding of the international policymaking capacities that need to be resourced by AfT initiatives. A key issue, for example, is the ability of developing countries and developing country experts to meaningfully participate in international standard-setting bodies. Discussions on standards are, for example, often the source of non-tariff barriers to trade and are of particular concern to developing countries when it comes to environmental goods and services. Such standards are often of great consequence to the developmental and environmental priorities of developing countries, including the prominent case of "eco-labels." For all effective purposes, standards are competitive tools and shape supply chains. An inability to participate in the development and deployment (or governance) of standards can only exacerbate existing dependencies. Yet, the level and quality of participation by developing countries in such deliberations leave much to be desired. Their ability to fully participate in international standard-setting is further restricted by the fact that much of this standard-setting is voluntary and emerges either from private sector initiatives or from domestic policy debates within industrialised countries. Standard-setting is one example, but not the only one, of why AfT investments need to broaden the development of capacities in developing countries (Rotherham, 2007).

5.3.3.2 Domestic implementation capacities Equally important for developing countries are the issues related to domestic implementation capacities. These can be related to institutional as well as infrastructural constraints and will require AfT initiatives for an assessment of both types of investments. As more and more trade policy matters focus on environmental dimensions to its implementation and as more and more of environmental policy is now seen to have trade implications, there is a growing, but often unfulfilled, need for strong institutional linkages between environmental and commerce-related implementing agencies. For example, in the area of illegal trade of natural resources and endangered species, it is often the case that environmental and policing agencies are each under-resourced in their own rights and, moreover, have few if any resources for coordinated action. The capacity deficits in such a case are institutional as well as infrastructural. A very important issue in terms of domestic implementation capacities in the context of AfT is that of Environmental Impact Assessments (EIAs) and Strategic Environmental Assessments (SEAs) (Abaza and Hemwey, 2001). Because much of AfT is likely to be used for infrastructure projects, it will be important to ensure that relevant environmental and integrated assessments are used as tools to integrate environmental considerations in projects, policies and plans. This, in turn, would require that the capacities to conduct such integrated assessments are also resourced. Another example may be that of trade facilitation (Chaturverdi, 2007). A comparatively new area in trade policy, trade facilitation can have significant implementation relevance for environment and sustainable development, for example through transport-related issues. Cross-border transport of goods is not merely a trading concern but also a development and environment concern, such as related issues like labour movements and transport-related pollution. Here, again, a focus on enhancing implementation capacities is needed both institutionally and in terms of infrastructure.

5.3.3.3 Stakeholder participation capacities The third element is related to both of the above and concerns the creation of policy space for meaningful participation by various stakeholders, including non-state actors (Nijnkeu, 2007). As suggested above, a lack of public participation in policy usually leads to a lack of public buy-in and sometimes heightened public distrust towards the policy. However, the reason to invest in enhanced stakeholder participation is not simply to 'buy' "buy-in' and ownership, but also to facilitate a mechanism that can and does often lead to better and more implementable policies. Once again, the logic for handling environment-related trade issues in the context of sustainable development is particularly compelling. Environmental issues, particularly those that are most pressing from a sustainable development perspective and relate to the poorest and most marginalised of populations, tend to be extremely local, and often remote.

This, for example, would be particularly true in the case of intellectual property rights related to indigenous knowledge. Biosafety issues in such a context highlight the need for paying attention to broad stakeholder participation. Traditionally trained policymakers can often lack the expertise or relevant knowledge to fully understand the complications and challenges of implementation on such issues because the 'key partner' may not be commercial enterprises in this case but dispersed, small, poor farmers. Examples include fisheries, in some organic agriculture products, and some biological resources in many parts of the world (Kurien, 2007). Broadening this understanding of the identity of the relevant stakeholders and devising innovative mechanisms for ensuring stakeholder participation in trade policy implementation is likely to become an increasingly important challenge as the world moves ahead towards realisation of more and more linkages between trade and environment concerns, especially at the implementation stage (Gueneau, 2007).

5.4. FINAL THOUGHTS

The strategic approach that AfT laid out throughout this chapter has highlighted the need for collaboration and integration across sectors. Implementing such a strategy would require similar collaborations and a similar integration of common goals amongst various agencies. It would also require a pooling of resources and a recognition of the comparative strengths of each partner.

Although prepared by UNEP, the chapter has consciously rejected taking a narrow institutional perspective focused solely on the environment and has instead tried to craft a logic that encompasses the common goal of all involved agencies; i.e., sustainable development. There is an obvious institutional strength within UNEP on issues that relate to the environment and to the integrated analysis and assessment capacities that have been discussed above. These can be a central thrust for UNEP to facilitate a cross-institution strategy for implementation. However, realizing the ideas laid out here will require close collaboration with other institutions, including, for example UNCTAD, on the issues that relate to the specifics of trade concerns, and development institutions such as UNDP, FAO and the Regional Economic Commissions on the many deep linkages to social and economic issues that are as much the backbone of the sustainable development construct as the environment. The exact structure of these collaborations and division of labour need to be further explored jointly, but what should be clear from the paper is that implementing the type of cross-sectoral and integrated approach to AfT that is being advocated here will require, by necessity, a cross-agency approach to its realisation.

NOTES

1 In discussing the WTO's Global Review of AfT, Pascal Lamy, the WTO Director General, noted that: "Aid for Trade can turn possibility into reality."

2 Text of the WTO Marrakesh Agreement and the Declaration of the Doha Ministerial meeting, both available at the WTO website.

REFERENCES

Abaza, H. (2007). The Case for Integrated Assessment. In A. Najam (Ed.) et al., *Trade and Environment: A Resource Book* (pp. 19–22). Winnipeg, Canada: International Institute for Sustainable Development, 2007).

Abaza, H. and R. Hemwey (2001). Integrated Assessment as a Tool for Achieving Sustainable Trade Policies. *Environmental Impact Assessment Review*, 21(6), 481–510.

Banerjee, A.V. (2007). *Making Aid Work*. Cambridge, Massachusetts.: MIT Press.

Bayne, N. (2004). Hard and Soft Law in International Institutions: Compliments, Not Alternatives. In J.J. Kirton and M.J. Trebilcock, *Hard Choices, Soft Law: Voluntary Standards in Global Trade and Environmental Governance* (pp. 347–352). New York: Ashgate Publishing.

Chaturverdi, S. (2007). Putting the Environment into Trade Facilitation. In A. Najam, M. Halle and R. Meléndez-Ortiz (Eds.), *Trade and Environment: A Resource Book* (pp. 191–198). Winnipeg, Canada: International Institute for Sustainable Development.

Cline, W. (2004). *Trade Policy and Global Poverty*. Washington, DC: Peterson Institute.

Collier, P. (2007). *The Bottom Billion: Why the Poorest Countries are Failing and What Can be Done About It*. New York: Oxford University Press.

Dijkstra, G. (2002). *Programme Aid and Development: Beyond Conditionality*. New York: Routledge.

Genee, O. (2007). Policy Coherence. In A. Najam, M. Halle and R. Meléndez-Ortiz (Eds.), *Trade and Environment: A Resource Book* (pp. 159–160). Winnipeg, Canada: International Institute for Sustainable Development.

Gueneau, S. (2007). New Policy Coherence Challenges. In A. Najam, M. Halle and R. Meléndez-Ortiz (Eds.), *Trade and Environment: A Resource Book* (pp. 165–168). Winnipeg, Canada: International Institute for Sustainable Development.

Halle, M. (2005). Where *Are We in the Doha Round?*. Winnipeg, Canada: International Institute for Sustainable Development.

Halle, M. and R. Meléndez-Ortiz (2007). The Case for a Positive Southern Agenda on Trade and Environment. In A. Najam, M. Halle and R. Meléndez-Ortiz (Eds.), *Envisioning a sustainable development agenda for trade and environment*. New York: Palgrave Macmillan.

Heydon, K. (2006). After the Hong Kong Ministerial Meeting: What is at Stake?, OECD Trade Policy Working Paper No. 27, Paris.

Hoekman, B. (2002). Strengthening the Global Trade Architecture for Development: The post-Doha Agenda. *World Trade Review*, No. 1, 23–45.

Hoekman, B. and S. Prowse (2005). Economic Policy Responses to Preference Erosion: From Trade as Aid to Aid for Trade, World Bank Policy Research Working Paper 3721, Washington, DC.

Kurien, J. (2007). Fisheries Subsidies and Beyond. In A. Najam (Ed.) et al., *Trade and Environment: A Resource Book* (pp. 103–106). Winnipeg, Canada: International Institute for Sustainable Development.

Nadal, A. (2007). Redesigning the World Trading System for Sustainable Development: A View from Central America. In A. Najam (Ed.) et al., *Envisioning a Sustainable Development Agenda for Trade and Environment*. New York: Palgrave Macmillan.

Najam, A. (2004). Trade and Environment Negotiations after Doha: Southern Priorities and Options. In Sustainable Development Policy Institute, Pakistan, *Sustainable Development: Bridging the Research/Policy Gaps in Southern Contexts*. Karachi: Oxford University Press.

Najam, A. and N. Robins (2001). Seizing the Future: The South, Sustainable Development and International Trade, *International Affairs*, 77(1), 93–111.

Najam, A., M. Halle, and R. Meléndez-Ortiz (2007). *Envisioning A Sustainable Development Agenda for Trade and Environment*. New York: Palgrave Macmillan.

Najam, A., M. Papa, and Taiyab, N. (2007). *Environment and Globalization: Five Propositions*. Winnipeg, Canada: International Institute for Sustainable Development.

Nijnkeu, D. (2007). *Aid for Trade and Development*. Winnipeg, Canada: International Institute for Sustainable Development.

Page, S. (2007). Aid for Trade: A New Issue in the WTO. In D. Nijnkeu (Ed.), *Aid for Trade and Development*. Cambridge: Cambridge University Press.

Riddell, R. (2007). *Does Foreign Aid Really Work?*. New York: Oxford University Press.

Rodrick, D. (2001). *The Global Governance of Trade as if Development Really Mattered*. Cambridge, Massachusetts: Harvard University/ New York: UNDP.

Rotherham, T. (2007). Standards and Labelling. In A. Najam, M. Halle and R. Meléndez-Ortiz (Eds.), *Trade and Environment: A Resource Book* (pp. 179–180). Winnipeg, Canada: International Institute for Sustainable Development.

Rotmans, J. and M. van Asselt (2004). Integrated Assessment: A Child on its way to Maturity. *Climatic Change*, vol. 34, pp. 3–4.

Tarp, F. (2000). *Foreign Aid and Development: Lessons Learned and Directions for the Future*. London: Routledge.

UNCTAD (2007). The Potential Impact of the Trade for Aid Initiative, G-24 Discussion Paper Series No. 45, Geneva.

Wijayasiri, J. and S. Samad (2007). Preference Erosion and Aid for Trade: A South Asian Perspective, CUTS Briefing paper SAFIT-II 3/2007, Jaipur, India.

Wing Yee Lee, B. (2007). Promoting Policy Coherence. In A. Najam, M. Halle and R. Meléndez-Ortiz (Eds.), *Trade and Environment: A Resource Book* (pp. 161–164). Winnipeg, Canada: International Institute for Sustainable Development.

WTO/OECD (2007). Aid for Trade at a Glance 2007: First Global Review, Geneva: WTO/OECD.

LORENZA JACHIA, ECE

CHAPTER 6

AID FOR TRADE: SUPPORTING THE USE OF STANDARDS

6.1. INTRODUCTION

The goal of the Aid for Trade (AfT) initiative is to nurture productive capabilities so that the opportunities of a more liberal trading system can be fully exploited by all countries. This chapter focuses on aid specifically geared towards standards, as one key component of the AfT envelope. It takes a broad look at technical assistance and capacity-building in the area of standards, encompassing assistance by different actors, and for a wide spectrum of related objectives. The term "standards" will be used to refer to technical regulations that are developed for reasons of health, safety and environmental protection (among other goals), as well as voluntary and private standards that are designed to promote the smooth functioning of supply chains.

With traditional barriers to trade – such as tariffs and quotas – being progressively eliminated, the ability to comply with technical regulations and use international standards emerges as a key factor of success on national and international markets. The chapter argues for increasing resources for assistance in trade-related standardisation matters, and for resources being used to envision and implement tailor-made, coherent and integrated strategies that will enable the participation of developing countries and countries with economies in transition as full players in the standards-development process.

Participating as full players means identifying and defending national priorities with a full understanding of the terms of the debate, both within the relevant standard-setting institutions, and in the work underway in the WTO. Such effective participation requires expertise developed at a local level, through standards use, compliance with technical regulations, and research and development. Effective coordination among stakeholders is also needed so that expertise developed nationally through these activities translates into informed and effective participation at a regional and international or multilateral level.

Currently, the thrust of assistance in the area of trade-related standards is aimed at increasing compliance through the upgrading of the relevant infrastructure – such as metrology and conformity assessment laboratories – and human resource development. While this is valid, it needs to be recognised that compliance with regulations, standards' use and participation in international standards-setting activities and WTO work are intrinsically linked, and projects in all these related areas should be developed as part of a coherent whole.

Lorenza Jachia wishes to thank the UNIDO Office in Geneva for the helpful comments provided.

P. De Lombaerde, L. Puri (eds.), *Aid for Trade: Global and Regional Perspectives*,
United Nations University Series on Regionalism 2,
DOI 10.1007/978-1-4020-9455-2_7,

In more detail, the idea the chapter exposes is that assistance in the area of trade-related standards, needs to focus on three priority areas:

- Strengthening the participation of developing countries and countries with economies in transition in the standards-related work underway in this area in the WTO as well as in the specialised technical institutions;
- Assisting firms and institutions as they respond to a growing pressure to comply with safety, quality and technical regulations of increasing complexity;
- Promoting the use of standards as a means of making firms more productive and helping them move up the value chain in their own markets.

The first area of priority engages institutions as the key players, while the second two are more specifically directed to the business community. At the same time, the involvement of all the stakeholders across the board is a key factor of success, because the input of business is needed to identify a country's priorities in international standards negotiations, and because business cannot successfully compete in heavily regulated markets without high-quality public infrastructure. Assistance is also needed in the overall coordination of work underway in standardisation matters at a national and regional level.

The plan of the chapter is as follows. Section 6.2 makes a case for increasing effective participation and promoting stakeholders' balance in standards-setting activities as well as in the work of the WTO SPS and TBT Committees. Sections 6.3 and 6.4 discuss how business can benefit from compliance with technical regulations as well as an increased use of standards. Section 6.5 describes on-going and completed technical assistance and capacity-building projects in the field of trade-related standards. This is based on data available in the WTO Trade Capacity-building Database and a review of efforts to ensure consistency between needs and disbursements, as well as in the approach used by various executing agencies involved. Sections 6.6 and 6.7 summarise and draw policy conclusions.

6.2. MOBILIZING STAKEHOLDERS FOR EFFECTIVE PARTICIPATION

It is essential that there be effective participation in the regional and international work on trade-related standards during two crucial phases: (i) in the negotiation of new standards and the amendment of existing ones; and (ii) in the monitoring of the use of technical regulations to ensure against protectionist intent.

6.2.1 Effective Participation in the WTO Committees on Sanitary and Phytosanitary Measures (SPS) and on Technical Barriers to Trade (TBT)

The TBT and the SPS committees of the WTO are mandated to work to avoid unnecessary obstacles to trade from standardisation-related activities and sanitary and phytosanitary measures. The participation of developing countries and countries with economies in transition in these committees is essential to ensure that the agreements are implemented effectively and bring tangible benefits to the business community.

Establishing enquiry points and national notification authorities in all WTO member countries remains a priority. These provide information about domestic regulatory developments to the WTO membership and also relay to national stakeholders information pertaining to new regulations introduced by trading partners.

As of January 2008, and 13 years after the entry into force of the agreement, 18 countries had yet to establish a TBT enquiry point[1]. As regards SPS, 20 countries have yet to establish a national notification authority[2] and 12 an enquiry point[3].

Together the enquiry points and national notification authorities can be thought of as an entry point that will enable a country to participate in the work of the committees, at first "passively", by attending meetings and notifying relevant national measures, and progressively in a more active way, by putting forward expressions of concern and initiating dispute resolution procedures.

In the space of just a few years, both passive and active participation by developing and least developed countries in the work of the committees has greatly increased, albeit from a low basis[4]. To further increase participation and make it more effective a number of concerns need to be addressed.

First, as attending the meetings of the TBT and SPS Committees severely stretches the budgets of national Governments, financially supporting the attendance of delegates from developing countries and countries with economies in transition should remain a priority. However, assistance for this purpose, which falls under the aid-for-trade category of "Trade Policy and Regulations" (TPR), accounts for the smallest share of aid-for-trade flows, at merely 3 per cent. It even decreased 20 per cent during the period from 2002 to 2005 (OECD-WTO, 2007). More recently, however, and according to the joint WTO-OECD database, the annual growth rate over the baseline period is 3 per cent, reflecting assistance from the Trade Capacity-building Trust Funds (OECD-WTO, 2006).

Second, the large volume of notifications poses considerable challenges for the enquiry points and national notification authorities of developing countries and countries with economies in transition. For instance, 947 notifications were submitted in the period from 1 January to 28 September 2007, bringing the total number of SPS notifications submitted since the entry into force of the Agreement to 8,313 by October 2007 (WTO, 2007). To ease the strain that this implies for countries, the WTO secretariat has recently established the "SPS Information Management System", which allows users to track and obtain information about SPS measures in all member countries[5].

An interesting initiative has simultaneously been developed at a regional level by the Brazilian National Institute of Metrology, Standardisation and Industrial Quality (INMETRO). INMETRO operates two online information services for exporters: Using the "Solicitação de Informaçães" information service, exporters can obtain information about technical requirements their products have to comply with in foreign markets. And by subscribing to the *Alerta Exportador* e-mail notifications, they will receive early warning of notifications of new draft technical regulations and conformity-assessment procedures issued by WTO Member countries. More than a million Brazilian exporters currently use this service, which INMETRO provides free of charge to exporters in all four MERCOSUR countries. With sufficient funding, this Brazilian initiative could be replicated in other regions[6].

Regions such as the Pacific Islands that are very remote and find it costly and difficult to participate in these highly technical activities, have requested that a regional body assists them in the implementation of their TBT-SPS obligations.

A key area of concern for enquiry points is the lack of sufficient scientific and technological capacity to evaluate the potential impact of new measures for their domestic stakeholders. For example, enquiry points may have insufficient understanding about new hazards for which scientific expertise is predominantly based in developed countries. Or they may lack surveillance, toxicological and epidemiological data based on their own particular circumstances to challenge notifications of new SPS measures. Training and capacity-building are then essential for reinforcing participation by developing countries and countries with economies in transition. Making sure that enquiry points have sufficient time to respond to notifications – and that the notifications are translated into their working languages - are also important priorities (Henson et al., 2000).

Countries that are acceding to the WTO also benefit from assistance in implementing the requirements of the SPS and TBT agreements. An interesting extension of the projects underway would be to establish a body similar to an enquiry point in accession countries. This entity would be responsible for relaying information about regulatory changes in main export markets to national exporters.

6.2.2 Standards-setting Institutions: Representation of the Interests of Developing Countries and Countries with Economies in Transition

The work of the TBT and SPS committees hinges on standards that are developed in relevant regional and international standards-setting institutions including the Food and Agriculture Organisation (FAO), the Codex Alimentarius Commission, the United Nations Economic Commission for Europe (UNECE), the World Customs Organisation, the International Telecommunication Union (ITU), International Organisation for Standardisation (ISO) and the International Electrotechnical Commission (IEC). Of course, standards developed in these institutions have a significance that goes well beyond the WTO and trade-related aspects, and permeates all aspects of production and consumption of goods and services.

These standards strike a delicate balance between very different appreciations of what is necessary to guarantee quality and safety. A wide scope of stakeholders needs to be active in the standards-setting processes so that no one group makes its interest prevail. By participating effectively in these negotiations, representatives can expose the economic and technical capacity of developing country Members with respect to proposed standards and tilt the debate in their favour. Or, if a satisfactory compromise cannot be found, then representatives can plead for appropriate conditionalities to ensure a smooth transition. Understanding early on how the debate is evolving in one area also allows more time to devise and establish appropriate implementing strategies.

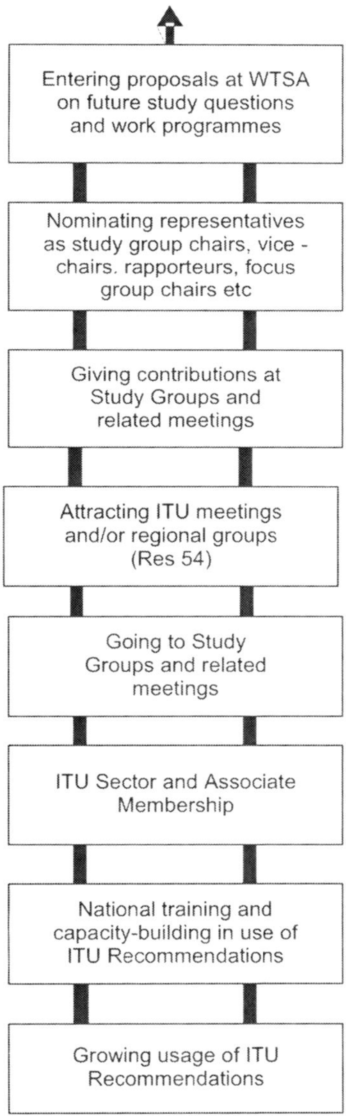

For these reasons, Aid for Trade should include as one of its elements the development of a coherent package that would include funding travel to international meetings of specialised standards-setting institutions; awareness-raising about the importance of trade-related standards; training for a core group of highly skilled professionals and promoting increased coordination among national stakeholders to define national interests and needs.

The Food and Agriculture Organisation, the World Health Organisation and the International Plant Protection Convention have established trust funds built around these priorities. ISO has established a technical assistance programme that includes training programmes and direct sponsorship of participation in meetings as well as the production of publications to guide experts joining technical committees[7]. More recently, ISO has also developed technical assistance and training activities that support all stakeholders in a balanced manner, including experts from developed countries, particularly in the framework of standardisation activities for social responsibility, through a special trust fund established for this purpose.

It is important to define "increased participation in standards-setting activities" as accurately as possible. ITU, for example, uses the "ladder" reproduced here to the left to represent how countries can – in practical terms – take increased responsibility in the development of standards in their organisation.

The first steps of the ladder are a growing usage of the existing recommendations and the organisation of national training workshops. This is followed by membership in ITU sectors and associate membership, as well as increased participation in meetings of regional study groups. One step up the ladder is to attract regulatory meetings to the country itself, allowing the local industry and regulators – not normally travel to attend these meetings – to obtain a more direct understanding of the standards-setting process as well as to make personal contacts with the secretariat and the working groups. These resources may be precious for local stakeholders to continue being updated on the negotiation process, even if they subsequently cannot attend on a regular basis.

In addition, if a national representation continues attending, it will be able to count on more informed inputs from the national stakeholders, even if they have only been able to attend only one session. The final steps are: giving contributions to study groups and related meetings, nominating representatives as chairs, vice-chairs and rapporteurs, and entering proposals at the World Telecommunication Standardisation Assemblies on future study questions (Kelly, 2007).

Similarly to ITU, ISO has elaborated on opportunities for exerting influence on its standards-development process. Again, a country will start by becoming more involved in the work of the Organisation – including by exploiting opportunities for twinning with more advanced partners and working with the policy development committees – then providing working party convenors, chairs and project leaders. Since no country can be expected to participate actively in all ISO committees on all subjects, it is imperative for every national member to select those projects in which it wishes to participate, on the basis of its national interests, and to develop a "bottom up", coordinated approach aiming to increase its representation in the project work (ISO, 2007).

UNECE is another body that develops international trade-related standards and best practice, with the active involvement of business:

- The United Nations Centre for Trade Facilitation and Electronic Business (UN/CEFACT) develops methods to facilitate national and international transactions, through simplifying and harmonising processes, procedures and information flows;
- The Working Party on Regulatory Cooperation and Standardisation Policies (WP.6) is leading the concept of the use of standards as the basis for technical regulation and is developing recommendations on a variety of policy matters relating to technical regulations, standardisation, conformity assessment, accreditation, market surveillance, quality/environmental management systems and metrology;
- The Working Party on Agricultural Quality Standards (WP.7) develops, interprets and promotes the practical application of international commercial quality standards for agricultural produce.[8]

Even though UNECE technical assistance activities focus on Central and Eastern Europe, the Caucasus and Central Asia, the standards and best practice developed at the UNECE are of a global nature. As one example of the significance of these standards, 36 of the UNECE commercial agricultural quality standards form the basis of European Union directives. They cover around 90 per cent of the market volume of fresh fruits and vegetable products traded throughout Europe. In practice, for purposes of quality, the EU accepts produce coming from non-EU countries if it is marked and controlled according to UNECE standards. UNECE needs to find more resources to support increased participation by developing countries in its standardisation activities and ensure better stakeholder balance. In doing so, it may build on the best practice developed within WTO, ITU and ISO described above, as well as experiences in other agencies.

6.3. INCREASING COMPLIANCE WITH TECHNICAL REGULATIONS TO PRESERVE AND ENHANCE MARKET ACCESS

Effective participation in standards-setting activities requires extensive experience in using the standards and in complying with the technical regulations applied in export markets in order to build awareness of the different options available, their implications and cost. More resources need to be drawn into standards implementation, on the one hand as a means to promote informed participation, and on the other to ensure smooth trade relations and avoid disruptions that are costly for both buyers and suppliers.

Border detentions represent one important dimension of this issue. These occur whenever testing by the importing country reveals that a product is not in conformity with the regulations in place; for example, because there is evidence of contamination from a toxic substance above the levels permitted.

Over the last few years, the number of detentions and rejections at border points have increased tremendously. Under the EU Rapid Alert System for Food and Feed (RASFF, 2006), the number of notifications rose almost tenfold from 698 in 1999 to 6,840 in 2006. In the United States, import refusals by the Food and Drug Administration (FDA) rose from 23,687 in 2002 to 77,260 in 2006[9]. Neither RASFF nor the FDA, however, records the value or the volume of the detained trade goods. One estimate of the value of agro-food trade affected by official product rejections for the year 2000/2001 was as high as $3.8 billion or 0.84 per cent of world exports (Jaffee and Henson, 2004).

The actual costs of import detentions are likely to be much higher than this estimate suggests. In fact, many of the costs incurred by the exporter cannot be recouped, over and above the value of the lost consignment. Furthermore, under the RASSF system, when one member country detains a single consignment for failure to comply with an SPS standard, the exporter will be included in a "rapid alert" list. Increased controls will be set in place in all EU Member States until a certain number of consignments have been cleared by the EU Member who initially imposed the alert. Furthermore, although other consignments may eventually not be detained, they will be significantly delayed and large costs will be incurred for import clearance. The combined effect of numerous import detentions – or of an import ban – can in some cases affect an entire export sector, as the example in Box 6.1. shows.

Box 6.1. Exports of fish from Lake Victoria

Starting in 1999, due to several cases of suspected fish poisoning and evidence of inadequate quality standards, the European Union imports of Nile perch fish from the region of Lake Victoria were banned. Uganda and Tanzania, the main exporters of Nile perch from the lake, suffered a tremendous loss, with fish exports dropping by more than 50 per cent as compared to the previous year (see chart below). A number of fish factories closed or operated under capacity, resulting in redundancy and unemployment. The United Nations Industrial Development Organisation (UNIDO) then set up a large technical cooperation project with a total budget of USD 4.6 million with financing from several bilateral donors: Austria, Denmark, Germany, Italy, Japan and the United Kingdom, as well as the United Nations Development Programme. The project established a solid foundation for fish safety so that in 2000, the EU ban could be lifted and new markets opened up, in particular in the United States.[10]

Imports of fish from Tanzania and Uganda

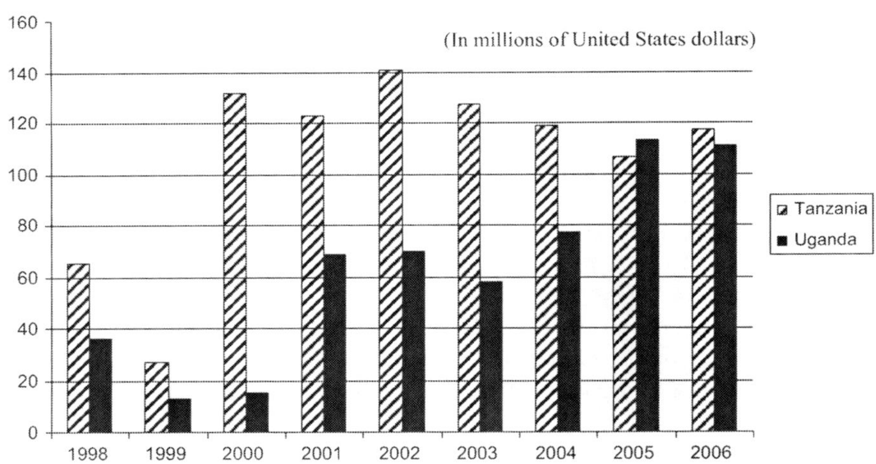

(*Source*: Eurostat)

6.3.1 Challenges at the Level of the Industry or Sector

Investing in compliance is a costly exercise. To provide an idea of the investments needed to restore and develop trade relations, Table 6.1. presents some summary information from case studies undertaken within a World Bank study programme

TABLE 6.1. Summary information from selected case-studies

Case	Reference	Costs of compliance (millions of United States dollars)	Exports (millions of United States dollars)	Costs as percentage of one year's exports
Senegalese fisheries sector	Niang (2005)	32.6 (cumulative total 1990–1999)	387.7 (average 1996–2001)	0.8
Bangladesh shrimp Sector	Cato and Lima dos Santos (2000)	18.1 (upgrade) 2.43 (annual maintenance)	225 (average 1996–1998)	1.1 (maintenance/ annual exports)
Indian spice industry	Jaffee et al. (2005)	14.5 (mid-1990s to 2003)	434 (2004–5)	3

Source: World Bank project on "Challenges and Opportunities Associated with International Agro-Food Standards"[11]

that focused on key commodities for which standards present a significant challenge to market access and competitiveness.

Expenditures incurred to restore trade relations when they have been disrupted typically amount to several million dollars, as the table shows. However, it is apparent that when the right strategy is in place, costs are recouped very rapidly, as also happened in the Lake Victoria case. The projects then become self-sustainable thanks to increasing export revenues from existing and new markets. It should be noted that the table shows cumulative, including a large proportion of non-recurrent, expenses made over the course of several years by a number of different actors, with substantial assistance from international agencies and bilateral donors. Obviously, without international assistance such a large amount of resources cannot be generated and sustained by developing countries and countries with economies in transition.

If we examine how the funds were allocated, we can make a broad distinction between costs that are incurred at the level of an industrial sector or a country, and those that are borne by single companies. In the first category, although there are wide variations from sector to sector, the main investments are directed at:

- Establishing or revamping laboratory facilities;
- Hiring and retaining specialised personnel to carry out tests;
- Investing in equipment to ensure cleaning/hygiene;
- Establishing or upgrading the institutional mechanisms or competent authorities;
- Revising the legal and regulatory framework and regulations;
- Upgrading the transport and storage facilities.

Traditionally, technical assistance projects place emphasis on the role of the Government and Government agencies in projects that aim at increasing compliance with trade-related standards. However, the business sector has a fundamentally important role in compliance, especially in developing and transition economies. For example, many companies or industry associations establish their own testing laboratories because the public administration cannot afford to equip and maintain them. Therefore, funds directed at public administrations should be complemented by assistance to the business sector.

6.3.2 Costs Borne by Single Companies

For any single company, complying with international standards and mandatory technical regulations is a costly business. The World Bank has compiled a comprehensive dataset through a 2002 survey administered to 689 firms in 17 developing countries. The data are freely accessible through the web at a high level of desegregation and detail. The firms were, for instance, asked to assess the investments that they had made to meet the technical regulations in export markets. These investments were then grouped into six different categories as shown in Chart 6.1. Almost 40 per cent of the firms had to make additional investment in plants or equipment, and 30 per cent had to hire additional labour.[12]

An important item missing from the database are the expenses incurred for training staff in the implementation of the standards. Nevertheless the analyses reveal a high level of awareness and understanding of the potential effects of international standards. As documented also in the case studies reviewed above, private firms have shown commitment to working with their foreign partner to resolve the problems as they arise, and have also demonstrated their ability to join forces with their competitors at a national level to establish common facilities and exert pressure on local authorities and to remove bottlenecks.

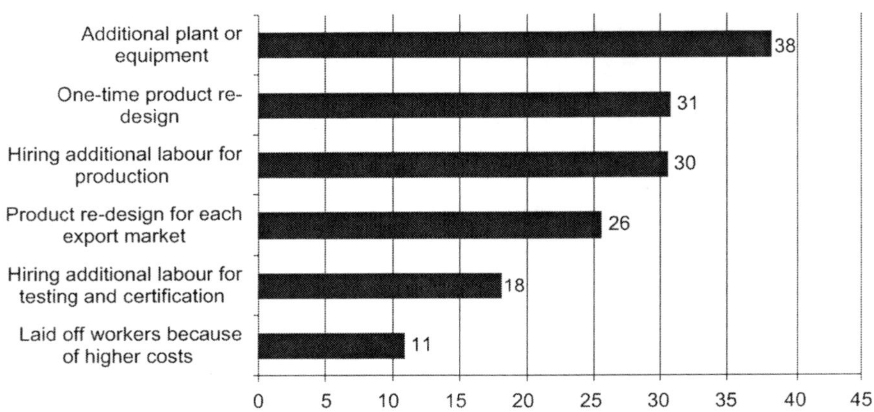

CHART 6.1. Percentage of firms that report additional investment in the following categories to comply with technical regulations on export markets (*Source*: Author calculations on World Bank Technical Barriers to Trade database)

Projects that aim at increasing standards use and compliance should be developed on the basis of surveys that document the needs of producers so that an informed choice, by the national authorities in consultation with all the stakeholders, can be made regarding the chosen standard or standards and the implementation strategies.

6.4. USING STANDARDS AT A COMPANY LEVEL TO INCREASE COMPETITIVENESS

Standards may often be seen as a means of complying with technical regulations and hence preserving or developing access to markets. Nonetheless, since they are developed by international experts and incorporate the latest research and know-how, they are also an important means to improve quality and reliability of goods and to climb the value chain to different and more lucrative niches.

For example, the family of standards known as "ISO 9000" requires "continual improvement of a firm's performance in the pursuit of excellence" in applying relevant regulatory requirements, producing in conformity to the customer's quality requirements, and enhancing customer satisfaction. Certification of a product with "ISO 9001: 2000" requires not only that an external audit assess an extensive sample of the firm's sites, functions, products, services and processes, but also that the firm's staff are trained for continuing a process of "continual review and assessment, to verify that the system is working as it's supposed to, find out where it can improve and to correct or prevent problems identified"[13].

Certification with ISO 9001: 2000 has an intrinsic value. It promotes an organisational culture based on the critical assessment of the firm's performance by its own staff on a continuing basis. While this is particularly true of the 9000 family of standards (and similarly ISO 14001), it is true in general that bringing standards developed by internationally recognised technical experts to the firm production floor will always bring about change and challenge established production methods.

Of high relevance for developing countries and countries with economies in transition is the recent ISO 22000 food safety management system which sets requirements for organisations throughout the food chain. This standard combines the CODEX principles of Hazard Analysis and Critical Control Point with pre-requisite programmes such as Good Manufacturing Practice with the establishment of an overall safety management system.

The ability of firms to use product and process standards is also one important element in the choices the Transnational Corporations (TNCs) make regarding their first and second tier suppliers. Producers that have a record of consistent quality in their output and can be relied on to undertake conformity assessment and certification directly on their own premises or at agreed upon public inspection points will be able to enter into long-term and stable contractual relationships with large clients. These agreements typically involve not only regular and recurrent orders, at pre-agreed prices, but also on-site training and transfer of technological and organisational know-how. On the other hand, those producers that have a mixed record of compliance with international standards will often have no choice but to sell their produce on more volatile international markets, where the TNCs and other smaller clients conclude "on-spot" transactions to fill needs for peak-season orders or unexpected surges in demand (Giuliani et al., 2005; UNCTAD, 2007).

Finally, standards' use may contribute to creating valuable strategic partnerships with research institutions. This is an especially important point because achieving a "regulatory objective" may require different production processes depending on the technological or climatic conditions of the firm or the farm. For example, if the objective is to achieve a minimum contamination level on consignments of fruit, different production methods may be required depending on where the fruit is grown. The same is true, although it may seem less intuitive, for technical standards as well, because the technology that is used for a given purpose in more technologically advanced economies may not be readily adaptable or may be too costly to integrate. Therefore, more research activities are needed towards devising practical and cost-effective ways to meet regulatory objectives that take into account the concerns of developing countries and countries with economies in transition.

6.5. ASSISTANCE IN THE AREA OF TRADE-RELATED STANDARDS TO DATE

This section reviews technical assistance and capacity-building undertaken to date within the area of trade-related standards, as well as the various attempts that have been made at identifying unmet needs and setting the priorities for action.

The first Global Trade Review, which took place in November 2007 at the WTO, together with the regional events that preceded it, promoted a better understanding of what kind of assistance is currently being provided and whether it meets developing countries and transition economies' needs. Yearly donor commitments during the period 2002-2005 were estimated to average USD 21 billion per year (OECD-WTO, 2007). This global estimate includes support to traditional categories such as trade policy and regulations and trade development, as well as the promotion of productive capacities and the major category represented by the support to setting up and maintaining of economic infrastructure. The latter represents an addition to what was traditionally included within the "technical assistance and capacity-building" categories.

The WTO-OECD Trade Capacity Building Database provides a large volume of data on trade capacity-building assistance and is today the main source for analysis and information. It captures the monetary value of support provided and also includes a short description of the project goals and expected outcomes. The charts presented here have been produced on the basis of the database, and a few qualifications are needed to appreciate the reliability of the analyses that are developed below.

First, standards-related technical assistance and capacity-building are mainly captured in the database under the TBT and SPS labels. However, capacity-building in standardisation matters, as defined for the purposes of this paper, goes beyond SPS

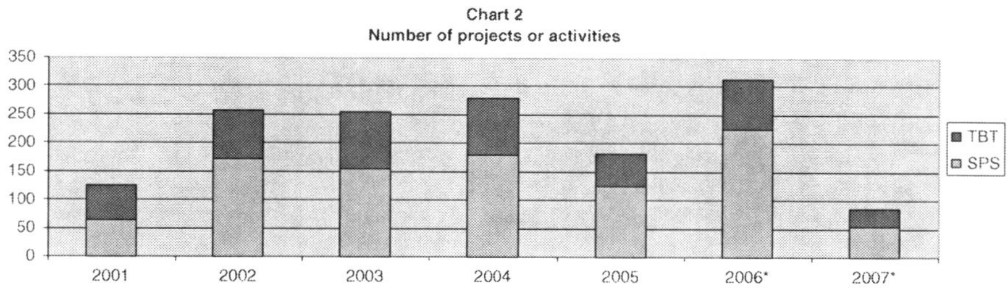

CHART 6.2. Number of project or activities. *Data for 2006 and 2007 incomplete (*Source*: WTO Trade Capacity-building Database)

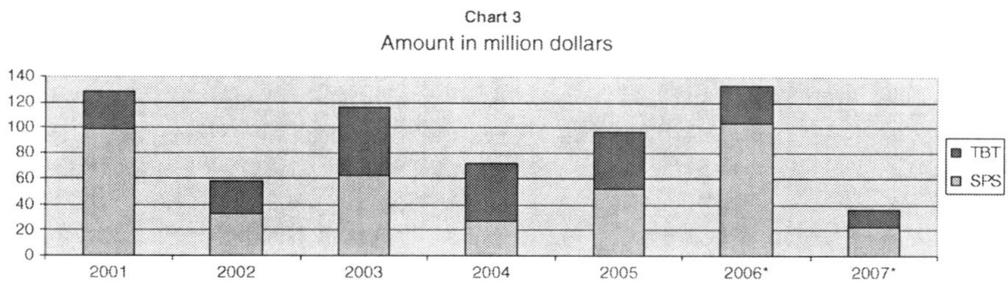

CHART 6.3. Amount in million dollars. *Data for 2006 and 2007 incomplete

and TBT. Supply-side productive capacity-building often involves assistance in standards implementation, in particular for quality and management systems standards. And indeed, the database records a number of projects within other categories, such as "Business Support Services and Institutions", "Trade Promotion Strategy Design and Implementation", "Market Analysis and Development", which provide assistance in standardisation matters. For coherence sake, it was however decided to include in the analysis only those projects that are listed within the SPS and TBT categories.

A second caveat is that the database is incomplete because not all donors and recipients fully disclose assistance projects. In particular, World Bank projects are not included in the database, although they play an important role in technical assistance in trade-related standardisation matters.[14]

The charts presented here attest to the limitations of the database. It is for example difficult to explain the variations in total donations from year to year, which are substantial both as regards the number of projects and their monetary value. It seems plausible that with improved collection and capture of data we would get a more coherent picture. Donors, recipients and implementing organisations should therefore be encouraged to report in full, and in as much detail as possible, the development assistance that is being made available.

At a global level, there are between 200 and 250 projects on issues related to TBT and SPS in a typical year. The value of the assistance provided ranges between USD 60 million and 120 million. SPS accounts for the majority of both the value and the number of the projects.

6.5.1 Assistance in the Area of TBT

Assistance in the TBT area aims at raising awareness and strengthening standards implementation capabilities, while standards development is supported mainly in terms of establishing standards information capacities, including the creation of enquiry points. In monetary value terms, a large share goes for building conformity-assessment infrastructure either by providing equipment for metrology or establishing testing and calibration laboratories.

The total number of projects listed in the database concerning TBTs is 490. Their average size is USD 530,000. The range among recorded projects is quite wide: 79 projects report very small grants (under USD 10,000), mostly to finance participation in training programmes. At the other end of the spectrum, 42 projects have a budget of between USD 1 million and 10 million, 3 have one of over 10 million dollars, the largest project being of 17 million.

The European Commission is financing or implementing almost all of the large projects. The share of projects implemented by international organisations is 18 per cent. The two agencies that are involved the most are UNIDO and ITC, followed by OECD, the regional commissions of the United Nations, ISO, UNDP, the World Customs Organisation (WCO) and FAO. The rest are implemented directly by bilateral agencies, and in isolated cases, by the beneficiary.

ISO activities in the area of TBTs and more in general in support of developing countries are reflected in its strategic framework entitled, "ISO Action Plan for developing countries". The objectives of the Action Plan are: raising stakeholders' awareness of the importance of standardisation, capacity-building, improving the use of electronic tools and strengthening the participation in technical work. The regional dimension is central to ISO work, which prioritises relations with regional and sub-regional standardisation bodies.

6.5.2 Assistance in the Area of SPS

The average size of the projects related to SPS measures, as recorded in the database, is USD 455,000. Out of a total of the 890 projects listed as active or having been active, 236 small projects report grants of below USD 10,000. At the other end of the spectrum, 57 projects have a budget of between one and ten million dollars, 6 projects one of over ten million, with the largest being of 42 million.

Again, almost all of the very large projects are financed and/or implemented by the European Commission. Similar to what was noted above for projects in the area of

TABLE 6.2. SPS and TBT projects as a percentage of the number of projects and their monetary value

	Number of projects		Monetary value of the projects	
	All trade policy and regulations projects	All trade policy and regulations, plus all trade development projects	All trade policy and regulations projects	All trade policy and regulations, plus all trade development projects
2001	7.8	3.5	19.7	5.9
2002	9.9	5.3	8.5	3.0
2003	10.2	5.0	12.9	4.0
2004	11.4	5.6	8.7	2.4
2005	7.7	3.3	10.5	3.1
2006*	10.5	4.5	10.9	3.8
2007*	10.7	7.8	9.9	7.1

Source: WTO trade capacity-building database.
*Data for 2006–2007 incomplete.

TBT, only 14 per cent of the projects are implemented by intergovernmental organisations, mainly UNIDO and FAO, followed by ITC, the World Organisation for Animal Health (OIE) and WHO.

Channeling more of the resources through intergovernmental organisations would ensure that these agencies' expertise could be relayed effectively. In particular, involving those agencies that are active in standards-setting in the standards implementation would allow for increased coordination among the stakeholders involved in the various stages of a standard's life. For example, unresolved issues that become apparent in the implementation of a standard could be brought to the attention of the experts when the standard is being revised. And agencies that are active in standards' implementation could usefully bring to projects' design the benefit of a perspective that includes the whole supply chain, therefore building up the policy coherence and credibility of the assistance programme[15].

Assistance in SPS and TBT is an important category as a percentage of the total assistance being reported in the TCBDB database. As shown in Table 6.2, these projects account on average for 10 per cent of the total number of projects relating to trade policy and regulations, and around 5 per cent of the total number of projects relating to all trade-policy regulations and trade development. In monetary value, the percentages are similar but fluctuate in a more pronounced way from year to year.

6.5.3 Regional Distribution of Assistance

The regional distribution of assistance is set out in Chart 6.4.[16] The low share of the Latin American countries in this distribution is not easy to account for, considering that these are relatively advanced economies that have a large industrial infrastructure

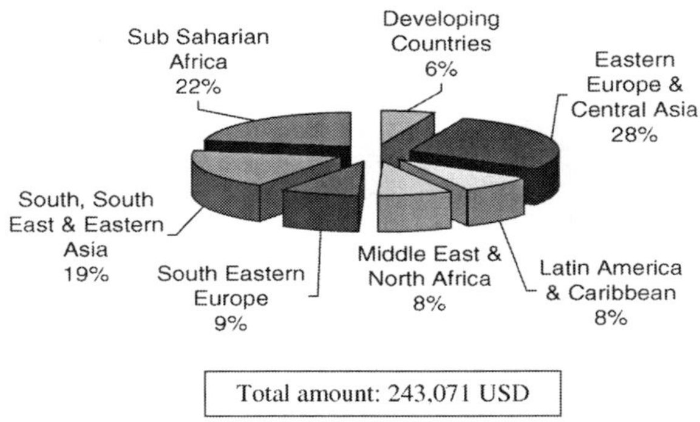

CHART 6.4. Percentage of total TBT donations by region

in place. On the other hand, the relatively high share of projects allocated to the countries of South Eastern Europe and Eastern Europe and Central Asia is clearly related to the exceptional assistance received by some of the countries in the region, and in particular those that acceded to the European Union in 2004, since a prerequisite for accession was the adoption and implementation of the *acquis communautaire*.

In addition, the countries with economies in transition clearly have special needs in the area of trade-related standards, because they started their transition to a market economy with limited resources in this area. Some of the countries had no national standards-related infrastructure, everything having been previously centralised in locations that – after the period between 1988 and 1992 – were no longer within the national borders.

Concerning the SPS donations, we see that the distribution is relatively more even, as each of the regions receiving more or less 15 per cent of the total amount, with the exception of South, South East and Eastern Asia that received 10 per cent and Middle East and North Africa that received 5 per cent. An explanation for this pattern may be that most developing countries export agri-food commodities and have articulated their needs more systematically in this area.

The regional dimension is of special importance as regards technical assistance in the area of standards. For example, because of the costs of the technical infrastructure needed to assess compliance with certain technical regulations, it may not be economically viable to set up national laboratories, and a regional approach might offer a feasible alternative.

Also, to further trade relations among regional partners, especially within South-South trade arrangements, increased regulatory cooperation in sectors of common interest brings a number of benefits, which are at the core of the work of the UNECE in this area.[17] The regional dimension of technical cooperation and capacity-building in the area of trade-related standards should therefore be maintained and strengthened. All the regional reviews have reaffirmed at a high and coordinated level their

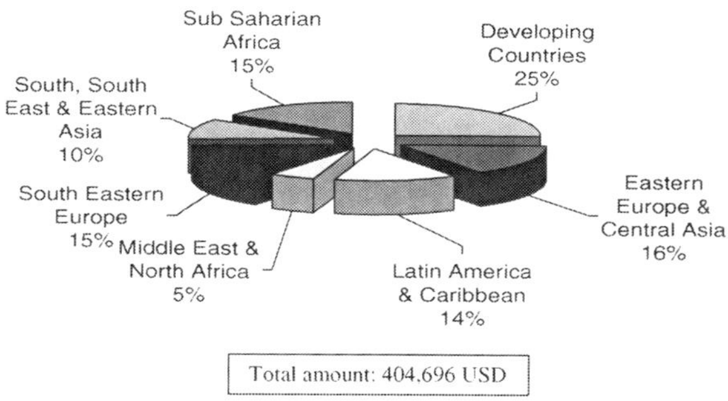

CHART 6.5. Percentage of total SPS donations by region

concern about increasingly stringent SPS and TBT measures in view of their weak capacities to meet international standards and assess conformity through testing, certification and accreditation, and have attracted attention to the need for increased assistance (WTO-IDB, 2007; WTO-ADB, 2007; WTO et al., 2007).

One important dimension, which was briefly touched upon in the beginning of this chapter, is the important contribution that "South-South" cooperation projects bring to building capacity in developing countries in the area of trade-related standards. These projects are often not based on financial contributions but instead aim at sharing the expertise and best practice that has been developed at a technical level through well designed and operational tools. It is very important that this assistance is appropriately accounted for and recognised, and that a repository of the expertise is developed in complementarity with the OECD-WTO trade capacity-building database.

6.6. THE WAY FORWARD: REINFORCING PROJECT OWNERSHIP BY BENEFICIARY AND FOSTERING COHERENCE AMONG EXECUTING AGENCIES

In recent years, a number of efforts have been directed towards assessing and prioritising needs in the area of standardisation. At the institutional level, the WTO-TBT Secretariat conducted a survey upon the Second Triennial Review, in 2002, which identified the following priorities, all of which remain relevant:

- Financial and technical support to establish conformity-assessment bodies and the relevant systems;
- Technical cooperation to strengthen and upgrade existing laboratories (e.g. through the provision of new equipment, training of staff and study visits);
- Assistance to purchase relevant international standards;

- Training in defining measurement uncertainty for calibration and test laboratories;
- Training in inspection activities and product certification by means of marks of conformity;
- Assistance to formulate a certification scheme that meets WTO requirements, and at the same time protects the interests of consumers and national industry (WTO, 2003).

More recently, the TBT Committee has also pointed to the need to facilitate the demand and supply of technical assistance and, in 2005, adopted a "Format for the Voluntary Notification of Specific Technical Assistance Needs and Responses". It appears, however, that this voluntary system is not yet being used sufficiently, which may hint to a need for technical assistance activities of an awareness-raising, needs-assessment and training nature (WTO, 2006).

6.6.1 Coordination Among Executing Agencies

Much of the development assistance provided over many years to build quality infrastructure has been fragmentary and has not been integrated into national strategies. Countries were receiving different and sometimes contradictory advice on how to set up their technical quality infrastructure effectively and efficiently. Responding to country demands requires coordination among the principal technical international organisations that have mandates to strengthen the technical infrastructures relating to metrology, standardisation and conformity assessment around the world.

In 2004, a first contribution towards increased coherence was made through the establishment of the Committee on Co-ordination of Assistance to Developing Countries in Metrology, Accreditation and Standardisation (JCDCMAS, 2004). The Committee members started by exchanging information and sharing experiences about the provision of technical assistance to developing countries.

The Committee went on to develop a common approach for providing of technical assistance for building technical infrastructure and for helping developing countries deal with a number of related challenges. It now recognises the urgent need for the coherent development of standardisation infrastructure. Since each element of the infrastructure is interdependent, a composite approach to standardisation, conformity assessment and accreditation is called for. For this, the Committee recommends that the following be considered:

- A thorough needs assessment of all parts of the economy;
- An appreciation that there is no ready-made infrastructure model to be slotted into place; Each country must create its own tailor-made solution;
- The technical infrastructure should be carefully developed and implemented to ensure sustainability, as there is no such thing as a "quick-fix";
- A clear statement of the resources and finance required should be prepared;

- National development of technical infrastructures does not preclude, but may well include, regional approaches, subject to the recognition of historical, political and cultural sensibilities.

In the area of SPS, a recent WTO survey of national notification authorities and enquiry points identified the following priorities for technical assistance:

- Raising awareness at the political level and among the public at large;
- Increasing coordination among different Ministries;
- Mobilising the relevant private-sector representatives;
- Encouraging regional and inter-governmental cooperation, including through mentoring and twinning;
- Assisting the enquiry points and national notification authorities in managing the inflow of notifications.

As a result of increasing pressure to improve coherence in responding to developing countries' needs in this area, in 2002 FAO, the World Bank, WHO and WTO established the "Standards and Trade Development Facility" to share information and support capacity-building for developing countries and countries with economies in transition in implementing SPS standards. This Facility has seen a rapid increase in the resources at its disposal, starting from USD 1 million in 2002 to USD 5 million in 2005. It is becoming the major clearinghouse as well as an important funds-mobilisation scheme.[18]

For least-developed countries, one importance source of funding for trade-related technical assistance and capacity-building in recent years is the "Integrated Framework for Trade-Related Technical Assistance to Least-developed Countries". This has, however, only marginally included standards and conformity-related issues. Projects mainly concentrated on raising awareness of the TBT and SPS agreements and supporting the establishment of enquiry points but failed. It failed to address at a more systematic level, the development of human resources and technical infrastructure in developing and transition economies so as to meet the challenges of potential barriers to trade and seize the opportunities of using standards as a vehicle to improve access to global markets.

While it is commendable to ensure coordination among executing agencies, care should be taken to avoid a proliferation of overlapping coordinating mechanisms. Coherence should be established among executing agencies not only at the operational level but also at the conceptual level, to ensure that a common message is consistently delivered, irrespectively of which agency is executing the project.

6.6.2 Strengthening Country Ownership

Along with coherence, the country ownership of projects developed in the area of SPS and TBTs also needs to be strengthened. In a review of projects related to SPS measures, a World Bank report notes that "a large proportion of assistance in this field by bilateral donors is driven by the "self-interest" or domestic considerations of the donors, such as ensuring that food imports are safe and preventing the spread of (..) pests and animal diseases from their main partners in the developing world"(Jaffee et al., 2005).

Similar considerations are true for the projects in the area of TBTs which often have as their main objective to provide officers in standards-setting bodies of the recipient country with information and training on the trade-related standards and practices of the donor country, with little or no consideration given to other approaches that may exist in other regions.

In sectors in which an agreed international standard has yet to be developed, assistance of this nature may, in some cases, be construed so as to influence international negotiations on trade-related standards in a way that is favourable to the donor country's interests. Additionally, if several donor countries act in a way that influences the development of standards in different regions in different ways, this may result in structural divergences in rule-making across regional blocs which will make an international agreement on a common standard much more difficult, if not impossible to achieve. Assistance in the area of to trade-related standards has effects that are similar to those of regional and bilateral trade agreements that contain provisions relating to rule-making. From this perspective both the regional trade agreements and the kind of development assistance that focuses exclusively on the donor's own regulatory approach – disregarding other approaches that may exist in other regions –may have important systemic effects on the development of common rules at a multilateral level (Isaac, 2006).[19]

Aside from these systemic implications, developing countries will benefit from projects that aim at assisting them in identifying their own regulatory needs and national priorities and guide them in choosing international standards that are adapted to their technological level.

For example, a project could fund a country team to engage in training and study tours of countries at a similar level of development, so as to choose among a variety of approaches those that are adapted to their specific conditions, and that can be more easily tailored to the requirements of different export markets. This would result in a level of ownership that will not be attained by simply adopting the regulations in use in the donor's country.

The above-mentioned World Bank report further states that projects related to SPS measures are typically "triggered by crises or imminent trade disruptions and involve reactive and remedial responses". The project regarding the export of fish from Lake Victoria described above in section 6.2 is one clear example.

Clearly, once a crisis is apparent or imminent, it can only be resolved through a technical assistance project and it is commendable that one is set in place. This should not, however, be the main source of technical assistance in this area. Crises are very costly for the exporting country, which will probably have suffered significant losses and damage to its reputation before the project is underway. Additionally, in facing a situation of crisis the donor may have to retain higher cost solutions to save time. In the long run, therefore, the emphasis should be on projects that prevent crises rather than aim at resolving them, including by promoting adherence by large importing countries to international standards.

6.7. CONCLUSIONS

As more of the scarce resources devoted to international development assistance are geared towards standards, it is important to define the priorities for action.

First, resources should be devoted to further debate about standards at a national level, including through comprehensive needs assessments, as well as mainstreaming standards in projects that focus on the development of productive capacities in specific sectors. Three broad areas of action can be identified:

- Reinforcing effective participation in standards-setting and in the relevant WTO institutions;
- Increasing compliance with technical regulations;
- Furthering the use of standards by business.

To reinforce participation, a policy priority should be for countries to secure funds not only for increased attendance but also for sustaining active involvement. Countries will need to be assisted in identifying national priorities, in consultation with all the stakeholders. The participation in standards-setting processes should be carefully defined and measured, including by defining "ladders" that document increased involvement, so as to give a visual and tangible dimension to progress made over time by national delegations.

The inclusion of new categories in the "Aid for Trade" database should not overshadow the need for continued assistance in traditional categories, and in particular in the category of "Trade Policy and Regulations", which appears instead increasingly marginalised, not least in the international debate.

In devising programmes directed at increasing compliance, particular attention should be given to involving the business sector, and its priorities and needs should be better documented and understood. The funds required to restore trade in the aftermath of a crisis are major, but are often recaptured in a relatively short span of time. Aid for trade should aim at preventing and not simply responding to crises situations.

Standards are not only a means of securing market access. They have a value in and of themselves because they are developed by international experts and embody the latest technology. They also raise output quality and provide indications to firms on how to better access markets and even to lower costs through the adoption of best practice. However, they also act as a selection device, assisting those producers who are able to adapt to them.

In reviewing the assistance that has been made available to date, it appears that the data are incomplete and do not allow a full appreciation of all the resources that have been invested in the area of assistance for trade-related standardisation.

In particular, the lack of a trend and the variation in the amounts registered from one year to the next are difficult to account for. Too few of the projects seem to be implemented directly by international organisations, despite the fact that these are the main repository of expertise in standardisation matters.

In recent years, attempts at coordinating the work of international organisations active in this area have multiplied and while this is doubtlessly beneficial, a proliferation of overlapping coordinating mechanisms is counterproductive. Coherence needs to be maintained not just at the operational level but also at the conceptual level to ensure that a common message is consistently delivered, irrespectively of which one agency is executing the project.

Together with coherence among implementing agencies, country ownership of projects developed in the area of SPS and TBTs needs to be strengthened. Too many of the projects seem to aim at facilitating the donor's imports from the country that benefits from the technical-assistance project. The assistance is not tailored to the needs and priorities of recipients. It also may have adverse systemic effects in those areas where a common regulatory approach has not been developed at a multilateral level, because it may result in a structural divergence in rule-making across regional blocs.

The real value added of aid for trade in the area of trade-related standards should be to empower recipient countries to make an informed choice regarding the instrument that best meets their development needs and the priorities of the national stakeholders. The assistance that is given in this context through "South-South" cooperation projects is of great value in this perspective and should be appropriately accounted for and recognised, including through the development of a repository of best practice that would represent a useful contribution to the OECD-WTO database.

In conclusion, comprehensive technical-assistance projects should be developed to assist developing countries and countries with economies in transition in establishing comprehensive national strategies in trade-related standardisation matters. Such projects should aim at putting in place effective coordination mechanisms among all the relevant stakeholders, including at the regional level if appropriate; ensuring active participation in standards-setting institutions in areas that are of key national interest; matching unmet needs for compliance with the expertise available in governmental and non-governmental organisations and fostering the use of standards by business.

NOTES

1 Of these, 10 countries are in Sub-Saharan Africa, 4 in Latin America and 4 in Asia.

2 Of these, 15 countries are in Sub-Saharan Africa, 1in the Middle East and North Africa and 4 in Asia

3 Eleven in Sub-Saharan Africa and one in Asia.

4 In 2003, high-income countries accounted for 19 per cent of WTO membership, but for 23 per cent of notification authorities, 21 per cent of enquiry points, 60 per cent of SPS notifications, 30 per cent of meeting participants, 55 per cent of the trade concerns raised, and 84 per cent of all SPS disputes. See Walkenhorst (2003).

5 http://spsims.wto.org/

[6] For more details please see: http://www.inmetro.gov.br/ This is but one example of the "South–South" cooperation projects implemented by Brazil, and more in general by emerging economies. These projects, which are again discussed below, bring the benefit of the expertise of relatively advanced developing countries to other developing and least developed countries, in the area of trade related standards as well as others.

[7] As another example, the Robert Oteng Fellowship is used to sponsor the studies of individuals from developing country national standards bodies. Projects undertaken under the auspices of the Fellowship support the development of human capacity in the field of standardisation.

[8] See: http://www.unece.org/trade/wp6, www.unece.org/cefact and http://www.unece.org/trade/agr/

[9] For the European Union, data are available at http://ec.europa.eu/food/food/rapidalert/index_en.htm, while for the United Sates at http://www.fda.gov/ora. RASFF is a network involving the European Commission, the European Food Safety Authority (EFSA), as well as the Member States of the EU and of the European Economic Area (EEA)

[10] See UNIDO "Trade Capacity-building: Case Studies: Lake Victoria" Available at www.unido.org

[11] The World Bank (2007). Permanent URL: http://go.worldbank.org/GHIT6173E0. See also, Jaffee et al. (2008).

[12] See: World Bank Technical Barriers to Trade database: http://go.worldbank.org/18YE2SZTJ0 (permanent URL).

[13] See www.iso.org and Wikipedia: ISO 9000.

[14] Most of the support provided by the World Bank in this area is through a sub-component of a broader programme, i.e. in export promotion and competitiveness, regulatory reforms, agricultural diversification, etc, making it difficult to extrapolate with precision the resources devoted to standards implementation.

[15] This paper cannot do justice to the UNIDO and the ITC technical assistance activities in this field. For more details about these two organisations programme of work, including their activities on trade-related standards please see, respectively: UNIDO (2007) and the UNIDO Trade Capacity-building website and ITC (2007) and the ITC website on "Aid for Trade".

[16] The category "Developing countries" in the WTO database refers to projects that have no pre-established geographic coverage.

[17] The UNECE Working Party on Technical Harmonisation and Standardisation Policies (Working Party 6) works to explore how national (mandatory) technical regulations could make wider use of international standards and to promote, where feasible, greater regulatory convergence in specific sectors, including telecommunications terminal equipment and earth-moving machinery.

[18] See: http://www.standardsfacility.org

[19] These issues have been explored by recent works that examine the role of rule-making in regional trade agreements.

REFERENCES

Cato, J.C. & Lima dos Santos, C.A. (2000). Costs to Upgrade the Bangladesh Frozen Shrimp Processing Sector to Adequate Technical and Sanitary Standards and to Maintain a HACCP Program. In L. Unnevehr (Ed.), *The Economics of HACCP: New Studies of Costs and Benefits* (pp. 385–402). St. Paul, MN: Eagan Press.

Giuliani, E., Pietrobelli, C. & Rabellotti, R. (2005).Upgrading in global value chains: Lessons from Latin American clusters, *World Development* 33(4), 549–573.

Henson S.J. et al. (2000), *Impact of sanitary and phytosanitary measures on developing countries*, Reading: University of Reading.

ISO (2007). *Joining In*, Geneva:ISO.

ITC (2007). The ITC role in Aid for Trade, Geneva: ITC.

Jaffee, S. & Henson, S. (2004). Standards and Agro-Food Exports from Developing Countries: Rebalancing the Debate, World Bank Policy Research Working Paper No. 3348. Available at SSRN: http://ssrn.com/abstract=610392.

Jaffee, S. et al.(2008). Challenges and Opportunities Associated with International Agro-Food Standards, available at http://go.worldbank.org/NZRMWQWX30 (permanent URL).

Jaffee, S. et al. (2005). Food Safety and Agricultural Health Standards: Challenges and Opportunities for Developing Country Exports, *World Bank Report* No. 31207.

JCDCMAS (2004). *Building corresponding technical infrastructures to support sustainable development and trade in developing countries and countries in transition*, available at: http://www.bipm.org/cc/JCDCMAS/Allowed/Background_documents/Background_paper_final.pdf

Isaac, G. (2006). The interaction between levels of rulemaking in international trade and investment: the case of SPS measures. In S. Woolcock (Ed.), *Trade and Investment Rulemaking* (pp. 51–77), Tokyo: United Nations University.

Kelly, T. (2007). Powerpoint presentation on "Bridging the ICT standardisation gap", UNECE Symposium on "Trade Rules, regulations, and standards: different levels of rulemaking and their impact", October 2007, available at: http://www.unece.org/trade/ct/ct_2007/symposium.htm#english.

Niang, P. N. (2005). Study of the Costs of Compliance with Export Standards in the Senegalese Fisheries Industry, *Agriculture and Rural Development Discussion Paper*, World Bank.

OECD-WTO (2006). *2006 Joint WTO/OECD Report on Trade-Related Technical Assistance and Capacity Building*, Paris: OECD.

OECD-WTO (2007). *Aid for Trade at a Glance 2007*, OECD, Paris.

The Rapid Alert System for Food and Feed (RASFF) (2006). *Annual Report 2006*, Available at: http://ec.europa.eu/food/food/rapidalert/index_en.htm.

UNCTAD (2007). *The Least Developed Countries Report 2007*, Geneva: UNCTAD.

UNIDO (2007). *A practical contribution to Aid for Trade*, Vienna: UNIDO.

The World Bank (2007). *Initiative to Mainstream Standards in Bank Operations.* http://go.worldbank.org/9NV6CDJPE0. Accessed at 9 November 2007.

WTO (2003). *An analysis of the priorities identified by developing country members in their responses to the questionnaire for a survey to assist developing countries to identify and prioritise their specific needs in the TBT field*, G/TBT/W/193 & G/TBT/W/178, Geneva: WTO.

WTO (2006). *Fourth triennial review of the operation and implementation of the agreement on technical barriers to trade under Article 15.4*, G/TBT/19, Geneva: WTO.

WTO (2007). *Report on the activities of the Committee on Sanitary and Phyto-Sanitary Standards*, G/L/842, 20 November 2007, Geneva: WTO.

WTO & ADB (2007). *Mobilising Aid-for-Trade: Focus Asia and the Pacific: Report and Recommendations*, Geneva: WTO.

WTO, AfDB & UNECA (2007). *Mobilising Aid for Trade: Focus Africa: Report and Recommendations*, Geneva: WTO.

WTO & IDB (2007). *Mobilising Aid for Trade: Focus Latin America and the Caribbean: Report and Recommendations*, Geneva: WTO. References niet ok

UNIDO

CHAPTER 7

BUILDING PRODUCTIVE CAPACITIES FOR TRADE COMPETITIVENESS: ECONOMIC PARTNERSHIP AGREEMENTS (EPAs) AND UNIDO's RESPONSE

7.1. INTRODUCTION

This chapter was initially prepared for the High-level Panel "Building productive capacities for trade competitiveness: Economic Partnership Agreements (EPAs)", which took place on 5 December 2007 during the 12th Session of the UNIDO General Conference. It intends to provide extensive background information and discussion material on the EU-ACP Economic Partnership Agreements.

It is divided into three sections: Section 7.2 includes an overview of the EU-ACP economic and trade cooperation including a situation analysis of the economic position of the ACP region. Special emphasis is put on the development component and the Aid for Trade initiatives. Section 7.3 entails a discussion of the opportunities and challenges of EPAs and proposes several adjustment measures on how to foster regional integration, strengthen productive and trade capacities and mitigate the effects from trade liberalisation of industrial products. Section 7.4 presents the response to EPAs from UNIDO, including its approach and the experience of past industrial upgrading and modernisation programmes.

7.2. OVERVIEW OF EU-ACP ECONOMIC AND TRADE COOPERATION

7.2.1 *What are Economic Partnership Agreements?*

History The EU and the ACP countries have a long lasting relationship, first formalised by the Yaoundé Convention in 1963. The membership of the association agreement was extended in 1975 in the Lomé Convention, which included an agreement for preferential, non-reciprocal tariffs to the ACPs. The four Lomé Conventions, signed every 5 years, focused on areas such as development aid, unilateral

This chapter has been prepared under the direction and supervision of Mohamed Lamine Dhaoui, Officer-in-charge of Industrial Policy & Private Sector Development Branch, PTCI PSD, UNIDO by Adrian Schöning, Consultant to UNIDO. Ms. Anne-Christine Peherstorfer of the UNIDO Liaison Office in Brussels, Mr. Christophe Yvetot, Mr. Francois d'Adesky, UNIDO's former representative in Brussels, Mr. Farrukh Alimdjanov and Ms. Clementine d'Oultremont, UNIDO, also provided helpful comments and valuable inputs for the chapter.

P. De Lombaerde, L. Puri (eds.), *Aid for Trade: Global and Regional Perspectives*,
United Nations University Series on Regionalism 2,
DOI 10.1007/978-1-4020-9455-2_8, © Springer Science+Business Media B.V. 2009

liberalisation of trade, promote and expedite the economic, cultural and social development of the ACP states, promote diversification, human rights, democracy and good governance.

After the creation of the World Trade Organisation (WTO) in 1995, there was a pressure to re-negotiate the agreements. This led to the signature of the *Cotonou Agreement* the 23 June 2000, agreed for a total of 20 years. The main objective of this agreement is to "*reduce and eventually eradicate poverty, consistent with the objectives of sustainable development and the gradual integration of the ACP countries into the world economy*"[1].

The objectives of the Economic and Trade Cooperation in Cotonou are to foster the "*smooth and gradual integration of the ACP States into the world economy, [...] thereby promoting their sustainable development and contributing to poverty eradication in the ACP countries*" and at "*enhancing the production, supply and trading capacity of the ACP countries as well as their capacity to attract investment*" by "*creating a new trading dynamic between the Parties" in order to enable "the ACP States to manage the challenges of globalisation [...] thereby facilitating their transition to the liberalised global economy*"[2].

This should have occured by means of replacing the trade chapters of the Cotonou Agreement and negotiating Economic Partnership Agreements (EPAs).

Why has this seen to be necessary? After 25 years of preferential access to the European market the ACP states did not manage to increase their export shares to the EU and the rest of the world. ACPs share in world exports fell from 3.4% in 1976 to 2.04% in 2006 while the share of the ACP in total EC imports fell from 6.7% in 1976 to 2.95% in 2006 despite unilateral and preferential market access[3] (see Figure 7.1).

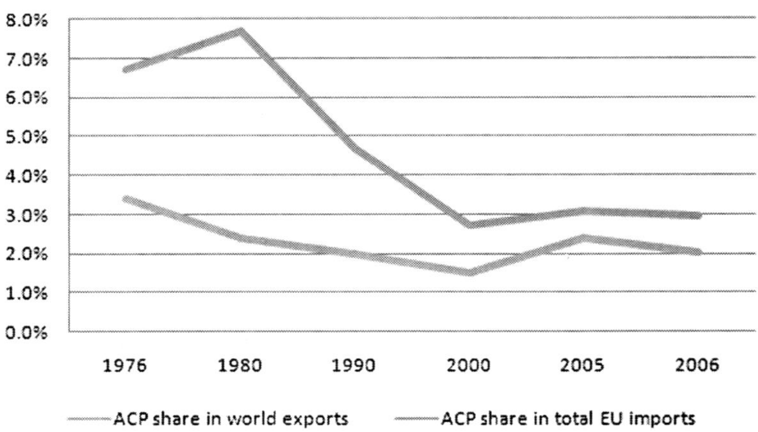

FIGURE 7.1. Economic marginalisation of ACP countries (*Source*: Data Eurostat)

The Lomé Conventions were also incompatible with WTO rules and discriminated against other developing countries. Article 1 of the General Agreement on Tariffs and Trade (GATT) concerning the general treatment of the Most Favoured Nation (MFN) clause says that "any advantage, favour, privilege or immunity granted by any contracting party to any product originating in or destined for any other country shall be accorded immediately and unconditionally to the like product originating in or destined for the territories of all other contracting parties."

The EU is allowed to give developing countries non-reciprocal trade preferences but the so-called "Enabling Clause" does not allow for discrimination between them, with exception to Least-Developed Countries (LDCs).

The current preferential market access is based on a WTO waiver obtained at the Doha Ministerial Conference on 14th November 2001, which, however, ran out at the end of 2007, requiring both parties to have put in place a WTO-compatible alternative. A further extension is not seen as likely, since other WTO members, especially the developing and emerging economies, are disadvantaged with the current situation and not willing to accept another waiver.

Economic Partnership Agreements Economic Partnership Agreements (EPAs) are the trade and development agreements the European Union is currently negotiating in parallel with six African, Caribbean and Pacific (ACP) regions (West Africa, East and Southern Africa (ESA), Central Africa, Southern Africa, the Caribbean and the Pacific). Full agreements are expected to cover trade in products (agricultural and industrial) as well as trade in services, economic and development cooperation, development finance cooperation, intellectual property rights, customs regimes, government procurement, investment and private sector development support, standards, technical regulations and conformity assessment, trade facilitation and competition policy.

EPAs lead to Free-Trade Areas (FTAs), which will progressively abolish "substantially all tariffs" between the Parties and will therefore conform to Article XXIV of the GATT. While the EU will completely open up its market from day one[4], ACP will progressively open regionally to each other, and then on a highly asymmetrical basis, to the EU. Former FTAs, for example between the EU and Morocco or South Africa, had a transitional period of 12 years. It is expected that substantially all trade will result for the six regions to liberalise 80 per cent or more of the total value of trade after a transitional period of up to 15 years. ACP will be able to exclude sensitive sectors and delay liberalisation to allow time for growing industries to adapt to change.

EPAs are based on four principles:

- Partnership: EPAs imply duties and privileges on both sides. The EU will remove non-trade barriers while ACP countries have to implement appropriate policies to tackle supply-side constraints and to reduce transaction costs.
- South-South integration: Regional integration lies at the heart of the agreement. If ACP regions are not able to reduce tariffs between themselves, facilitate trade,

resolve the problems arising from overlapping membership and pool together their resources, it is likely that there will be no integration of ACP countries into the world economy. The efforts undertaken in the last years have to continue on a meaningful basis and be reinforced. Special attention should be drawn to intra-regional trade diversion effects after complete liberalisation.

- Development: EPAs are instruments for development. Out of the 78 ACP countries, 41 are classified as Least Developed Countries. It has been stressed again and again by all parties[5], that the "development dimension" is central to EPAs in order to eventually achieve the objective of the Cotonou agreement: eradication of poverty.
- WTO-compatible: EPAs will build on rules of the WTO, and, if possible, go beyond the WTO. They will define bilateral trade relations intended to alleviate all practical barriers to trade between the two regions. Furthermore the negotiations are strengthening ACP capacities for further multilateral trade talks.

The main objectives can be summarised as follows:

- Create "North-South-South" agreements linking the world's largest customs union, the EU, with existing and upcoming customs unions in the South.
- The attainment of sustainable development, the elimination of poverty as well as a smooth and gradual integration into the world economy.
- The achievement of sustainable economic growth, expansion and modernisation of private sector, lowering unemployment levels as well as better access to productive resources.
- Achieve increased value-added of merchandise products to lower level of primary products, especially for agricultural products. Promote industrialisation by strengthening productive and trade capacity.
- Develop adequate infrastructure and tackle legal and administrative problems related to the export of goods and services to the EU. Further reduce non-tariff barriers.

However, most regions and countries do not see themselves in a position to sign full comprehensive EPAs, with the exception of the Caribbean. Therefore, it is likely that goods-mainly interim agreements will be signed with ESA, Pacific and Southern Africa. Negotiations will continue in order to complete a comprehensive EPA in 2008.

7.2.2 Regional Groupings

There are 76 ACP countries negotiating EPAs[6], divided into 6 regional groupings (see Table 7.1). Four of these groupings are in Sub-Saharan Africa, one in the Caribbean and one in Pacific. This division was not only made to save time and harmonise outcomes within a region but also because there are too many, sometimes too small

TABLE 7.1. Country groupings for EPA negotiations

	West Africa ECOWAS+ Mauritania	Central Africa CEMAC+ STP	East South Africa ESA	Sourthern Africa "SADC group"	Caribbean	
1	*Benin*	Cameroon	*Burundi*	*Angola*	Antigua, Barb	Cook Is.
2	*Burkina Faso*	*Central Africa*	*Comoros*	Botswana	Bahamas	Fed.Micron.
3	*Cape Verde*	*Chad*	*Djibouti*	*Lesotho*	Barbados	Fiji
4	*Gambia*	Congo	*Eritrea*	*Mozambique*	Belize	*Kiribati*
5	Ghana	*Congo (D.R.)*	*Ethiopia*	Namibia	Dominica	Marshall Is.
6	*Guinea*	*Equat. Guinea*	Kenya	Swaziland	Dom. Rep.	Nauru
7	*Guinea Bissao*	Gabon	*Malawi*	*Tanzania*	Grenada	Niue
8	Ivory Coast	*S. Tome, Princ.*	Mauritius	South Africa	Guyana	Palau
9	*Liberia*		*Madagascar*		*Haiti*	Papua N. G.
10	*Mali*		*Rwanda*		Jamaica	*Samoa*
11	*Mauritania*		Seychelles		St Lucia	*Solomon Is.*
12	*Niger*		*Sudan*		St Vincent	Tonga
13	Nigeria		*Uganda*		St.Ch. & Nevis	*Tuvalu*
14	*Senegal*		*Zambia*		Surinam	*Vanuatu*
15	*Sierra Leone*		Zambabwe		Trinidad & Tob.	
16	*Togo*					

Source: European Commission, LDCs in Italic

ACP countries who do not have enough negotiating capacities and are relatively weak in terms of their trading positions with the EU and the rest of the world.

Different countries face different development challenges. More than half of ACP are LDCs. Moreover the multitude of Small Island Development States (SIDS), land-locked countries and heavily indebted poor countries (HIPC) underlines the reason for a suboptimal development situation in the region. Therefore, EPAs take into consideration the level of development of each country and their socio-economic impact on them.

7.2.3 Situation analysis: Economic Position of ACPs

The European Union and the African, Caribbean and Pacific countries do not only vary in the number of countries, the population size and the surface area but more importantly regarding EPAs, in economic terms.

The EU has a Gross Domestic Product (GDP) 25 times higher than the ACP, with a per capita GDP of more than 37 times higher. Single negotiating regions achieve only between 0.1 and 1.2 per cent of EU's GDP. The demographic, economic and social differences between ACP and EU are depicted at a glance in Table 7.2 below.

The most important trade partners for the EU can be found in the USA, Russia and the emerging market China, whereas, ACPs share for import and export is only 3 per cent.

TABLE 7.2. Demographic, Economic and Social indicators for the EU and ACP

	European Union	Africa, Caribbean and Pacific
Member countries	27 countries[7]	78 countries[8]
Population (1st January 2005)	489 million (a)	728.5 million (excluded SA) (b)
Surface area	4 million km^2 (c)	25 million km^2 (b)
GDP (2006)	€ 10917 billion (a) € 22.325 per capita (a)	€ 435 billion (b) € 597 per capita (b)
Total Export (2006) (d)	$ 4.538 billion (EU-25)	$ 246 billion
Share in world trade (2006)	37,87%	2,06%
Average growth rate (2000-05)	11,9%	15,1%
Percentage of Intra-trade (2005)	66,5%	11%
Total Trade within the two regions (2006) (b)	EU-25 exports to ACP: € 34 billion	EU-25 imports from ACP: € 39.9 billion
Major export partners (2006) (b)	USA (23%) Switzerland (7,4%) Russia (6,2%), China (5,4%) ... **ACP Countries (2,9%)**	USA (33,8%) **EU (22,7%)** China (11,3%) Japan (3,3)
Major import partners (2006) (b)	China (14,2%) USA (13,1%) Russia (10,1) ... **ACP Countries (3,0%)**	**EU (24,7%)** USA (12,5%) China (8,5%) Korea (4,9%)
Foreign Direct Investment (2005) (d)	Inward flows: $ 422 billion Outward flows: $ 554 billion	Inward flows: $ 24 billion Outward flows: $ 0,85 billion
UNDP Human Development Index (2004) (e)	High	8 countries high[9] 25 countries medium, remaining low
Least Developed Countries (LDC)	None	41 (34 in Africa, six in the Pacific, one in the Caribbean)

Sources: (a) Europe in figures, Eurostat yearbook 2006-07, 20 February 2007.
(b) EUROSTAT or World Bank, as quoted in DG-TRADE, Fiche Pays ACP, 07 August 2007.
(c) Key facts about Europe and the Europeans, European Commission.
(d) UNCTAD Handbook of Statistics 2006-07.
(e) United Nations Development Programme (UNDP), Human Development report 2006.

For the ACP however, the EU is the most important import partner and the second largest destination in terms of export value. The EU represents 38 per cent of the world trade while ACP share is only 2 per cent. Other important partners are the USA and increasingly China.

Median Human Development in ACP countries is low, reflected by the fact that within the ACP regions secondary and tertiary education levels are generally low as well as government's health expenditure, leading to a low life expectancy.

Over the past four years real GDP growth in ACP was constantly over six per cent, resulting from rising productivity, domestic investment and government consumption. The latter has been possible through debt relief and higher oil revenues. But still more than half of ACP states are LDCs, reflected in a low GDP per capita. It fluctuates between regions and between countries. At constant prices from 2000, in 2006 it varies between $93 in Dem. Rep. Congo and almost $16000 in the Bahamas.[10] Twelve and eleven LDCs are in West African and ESA, respectively, resulting in the lowest median.

Agriculture, industries, manufacturing and services The composition of the GDP varies from region to region and from country to country (Figure 7.2).

Africa is the least developed region of the world in terms of industrialisation and the Caribbean and Pacific are generally not very industrially developed either. The ACP account for a negligible share of global industrial output. Industry (without manufacturing) is especially pronounced in West Africa, Central Africa and Southern Africa, resulting from petroleum oils and diamonds. It ranges between 2.7 per cent

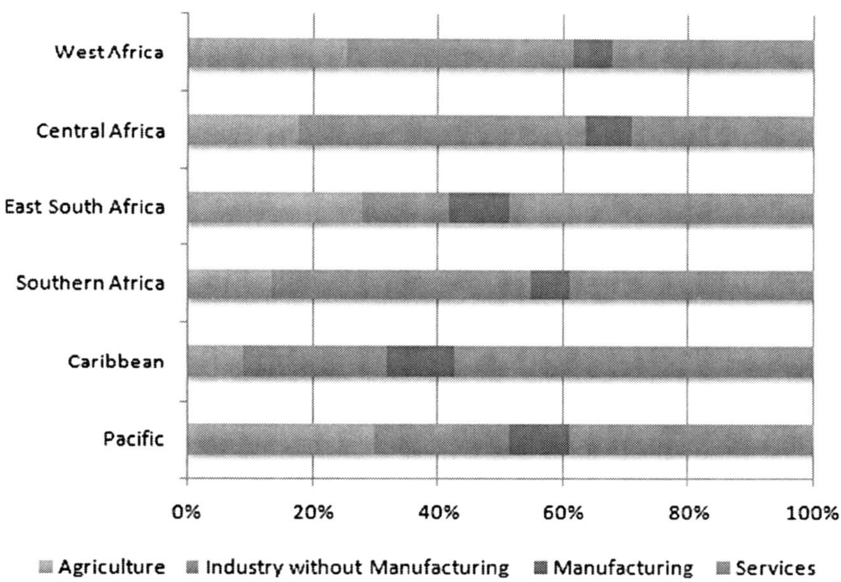

FIGURE 7.2. Share of GDP of different sectors (*Source*: Own calculations based on African Development Bank data)

in Zimbabwe and 95 per cent in Equatorial Guinea, whereas the median lies at 13 per cent. Industrialisation is a critical engine of economic growth and development, shown in the linkage between them.

The ACP countries are also marginalised in global manufacturing. Only ten ACP countries have already achieved an upper industrial stage with ratios above 15 per cent.

While the ACP countries account for approximately 9 per cent of the world population, its exports in the world accounted only for 2 per cent. The EU made up almost 38 per cent. These percentages correspond to $246 billion and $4.538 billion of global exports respectively (Figure 7.3). Both the EU and ACP had sustainable export growth rates since 2000, arriving at annually 11.9 per cent and 15.1 per cent respectively.

EU-ACP Trade Focusing on EU-ACP trade, in 2006 European Union imports from ACP countries amounted to almost €40 billion, corresponding to 2.95 per cent of total EU imports (Table 7.3). Primary products made up an impressive 78.8 per cent, of which 24 per cent in agricultural products and 43 per cent in energy. Manufactured products amounted to 20 per cent of total imports from ACP countries. This equals 1 per cent of total EU manufacturing imports. Between 2002 and 2006, imports rose by more than €9 billion, almost exclusively thanks to energy-related products.

In Figure 7.3, the main imported ACP products are depicted graphically. Food and live animals consisted mainly of cocoa beans, sugar, coffee, fish and bananas. Manufacturing products were transport equipment, textiles and clothes, beverages and tobacco and machinery. Spain, France and the United Kingdom are the EU countries that import most from ACP countries.

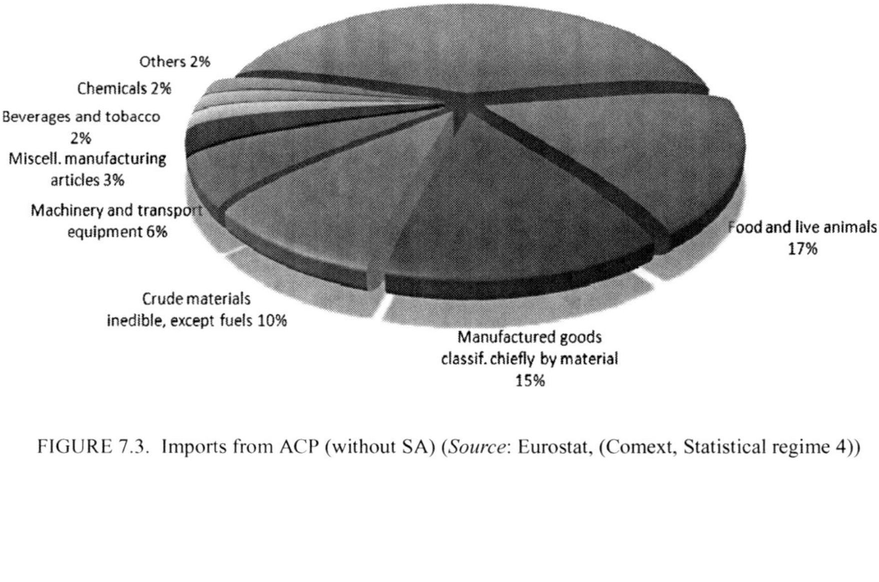

FIGURE 7.3. Imports from ACP (without SA) (*Source*: Eurostat, (Comext, Statistical regime 4))

TABLE 7.3. European Union imports from ACP countries in 2006

European Union, Imports from ACP	2006	%	Share of total EU imports
Total	39.851	100, 0%	2, 95%
Primary Products	31.392	78,8%	6,21%
of which:			
Agricultural products	10.120	25,4%	11,10%
Energy	17.339	43,5%	5,19%
Manufacturing Products	8.129	20,4%	1,0%
of which:			
Machinery	300	0,8%	0,10%
Transport equipment	1.933	4,9%	2,0%
Chemicals	615	1,5%	57,0%
Textiles and clothes	831	2,1%	1,01%
Beverages and Tobacco	615	1,5%	10,2%
Office/telecom. Equipm.	113	0,06%	0,3%

Source: Eurostat (COMEXT, Statistical regime 4)

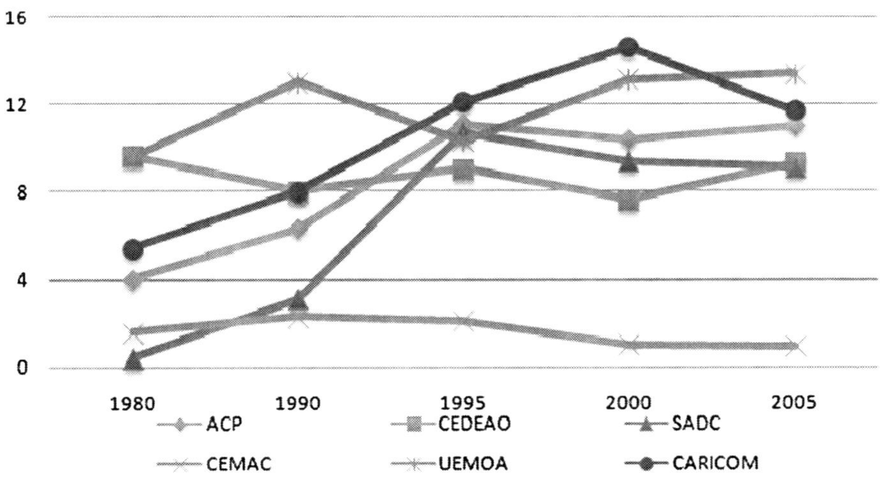

FIGURE 7.4. Intra-regional trade of EPA regional communities (*Source:* UNCTAD Handbook of Statistics 2006–2007)

Several ACP countries have been able to increase their export quantities to the EU during the last decades, facilitated by the Yaoundé and Lomé conventions. But even though, this was dependent on a few basic commodities, whose non-oil prices are in long-term decline, and a handful ACP countries. In 2006, 6 countries represented over 50 per cent of total EU imports from ACP, these being Nigeria (26% of ACP exports), Cameroon (7%), Angola (5%), Côte d'Ivoire (6%), Equatorial Guinea (4%) and Ghana (3%).

The destination shares of Sub-Saharan African (SSA) exports, including 4 out of the 6 EPA regional groupings, changed during the last decade. In 2001, the EU was by far the first largest recipient of their exports, receiving almost 35 per cent of total SSA exports, while in 2005 it decreased to 25 per cent[11]. Asia's share has been rising during the last years as well as exports going to the United States. This is due to increasing exports of fuels and textiles to the US and because of implementation of the AGOA scheme.

The social, economic and political gap between these two regions offer challenges to ACP countries, when trade should be liberalised vis-à-vis the European Union. On the other side, the region must take advantage of new opportunities offered by EPAs and globalisation by increasing capacities of enterprises, putting in place policies for upgrading, developing human resources, strengthening capacities for innovation, and, by accepting the facts of industrialisation. The agreements offer a chance to integrate the region into the world market and reduce poverty.

7.2.4 Development Component and Aid for Trade

Pro-development The EC states that "EPAs are above all instruments for development"[12]. Promoting sustainable development and contributing to poverty eradication by integrating ACP states into both their respective RECs and the world market, is one of the three main objectives of EPAs. Non-governmental Organisations (NGOs) like Oxfam often criticised the EPA process on the ground that the developmental issue has not been sufficiently addressed, fearing that the agreements instead of being conducive to development in the ACP countries, would rather be detrimental for the people living there (Oxfam, 2006).

Non-tariff barriers like Rules of Origin and SPS and TBT requirements, as well as trade regulations, accompanying measures and policies, and, effective processes for support delivery have to be addressed carefully. Adjustment costs and increased export volumes of European enterprises will need special attention in order to reach a pro-development outcome of EPAs. Trade regulations have to consider the different and specific needs of the regional groupings. Depending on their size, population and production sectors, the EU has to differentiate their trade-related actions.

If uncompetitive firms will be squeezed out of the market, further unemployment will be created. For these people, a safety net has to be created. Human resources development is necessary in order to find new working opportunities.

Beyond EPAs and their possible consequences, other elements have to be taken into account. Weaknesses at all levels prevent countries and their producers to export more and at a higher revenue. At the macro level, the lack of infrastructure and inefficient institutions is hindering fast and favourable trade and increases trading costs. At the micro level, producers sometimes do not know how to produce competitively or where and how to sell their products at a good price. Weak productive capacities, old trade-related policies, burdensome custom procedures, difficult compliance with SPS and TBT and restrictive Rules of Origin are all export constraints which will require special attention.

ACP countries have to overcome these constraints that deter them from taking full advantage of new market access opportunities to the EU and help them to integrate into global markets. Implemented in a multi-faceted approach, trade related technical assistance could bring sustainable growth and lead to poverty reduction.

Aid for Trade will be needed to achieve these stated objectives. Policies and accompanying measures have to be put in place in order to facilitate the implementation and mitigate the negative fiscal impact of the EPA. Long-term commitments for trade-related technical assistance (TRTA), based on specific needs assessments, are the key to ensure that benefits can reach the most vulnerable.

Aid for Trade Between 2002 and 2005 total Aid for Trade donor commitments reached $21 billion per year or 34 per cent average share in total sector aid[13]. This figure includes:

- $0.6 billion for *Trade policy and regulations;*
- $11.2 billion to build economic *infrastructure;* and
- $8.9 billion to promote *productive capacities* (including $2 billion for *Trade Development*).

Aid for Trade is a key complement to, but not a substitute, for further trade liberalisation under the EPAs. Increasing supply-side capacity and removal of structural weaknesses is key for a successful EPA implementation. This has also been reaffirmed by the European Commission in a recent communication[14]. EC's overall trade-related assistance between 2001 and 2005 amounted to €1.619 million for ACP countries[15].

Aid for Trade has been a controversial issue in the EPA negotiations. The 6 regional economic communities wanted to see trade-related assistance included in the negotiations, while the EC clearly wanted to separate market access negotiations from additional aid commitments. The reason was that the Cotonou Agreement foresaw only trade and trade-related issues to be discussed in the framework of EPAs and that a negotiation of policies should not turn into a negotiation over trade flows.

The main financial mechanism for ACP countries is the European Development Fund (EDF). The 10th EDF amounts to €22.6 billion for this period up from €15.2 billion for the 9th EDF. According to the EC a huge part of the substantial increase will be used for EPA related projects and programmes. The following budgets for support at the regional level under the Regional Indicative Programmes (RIPs) have already been allocated under the 10th EDFs (2008–2013):

- West Africa: €478 million (250% increase over the 9th EDF);
- Central Africa (CEMAC): €127 million (100% increase over the 9th EDF);
- East South Africa (ESA): €465 million (108% increase over 9th EDF);
- Southern Africa (SADC): €135 million (reduction from 14 to 7 beneficiary countries – amounts to a 33% increase over 9th EDF for the 7 countries);
- Caribbean: €132 million (93% increase over 9th EDF);
- Pacific: €76 million (130% over 9th EDF).

Since other development programmes concerning health, water, rural development and other areas are addressed under the EDF, civil society organisations fear that the strengthening of trade-related technical assistance may lead to neglect other priority areas.

ACP countries will have to struggle with its implications when they come into force. Therefore, the EC and its Member States strive to increase their volumes of EU Aid for Trade to €2 billion a year by the year 2010. (The amount of €1 billion will be pledged by the EC, €1 billion by the 27 Member States (additional to 10th EDF). The percentage reserved to support the ACPs should be in the range of 50 per cent.

It is difficult to assess if the amount of Aid for Trade pledged by the EC and their Member States will be sufficient to address all adjustment costs related to the implementation of EPAs. Costs will vary between regions and furthermore between different ACP countries. A study conducted by Milner[16] and commissioned by the Commonwealth Secretariat shows that the total EPA costs of adjustment would amount to €9.2 billion, including:

- €3.3 billion for fiscal adjustment support;
- €2.1 billion for trade facilitation/ export promotion support;
- €1.5 billion for production and em-ployment adjustment assistance; and
- €2.3 billion for human resources development and productivity enhancement support.

This study takes into account the degree of adjustment that countries will have to undertake and the difference in country size. Based on existing adjustment programmes being financed by the World Bank, the financial requirements arising from the design and implementation of mitigating measures are then calculated.

At the regional level the costs are distributed as set out in Table 7.4 below.

In addition to EU funds, other funding mechanisms in support of global Aid for Trade initiatives are in development, such as the Integrated Framework[17] (IF) and the WTO Aid for Trade Initiative, and should help ACP countries to address EPA challenges.

From draft EU/ACP EPA legal texts[18] emerge that the EU recognises the importance of cooperation in industrial development and competitiveness with the objectives of

- establishing, developing, restructuring and modernising a regions' industry and
- establishing an environment favourable to the development of private enterprise in order to stimulate the growth and diversification of industrial production.

Areas of cooperation should therefore include:

- Promote the development of activities in the areas of production, processing, marketing, distribution, financing, and development of commodities;
- Promote collaborations in Research & Development and encourage knowledge and technology transfers;
- Promote and strengthen innovation, diversification and value addition product development and quality;
- Strengthen the financial system in order to reduce the high transaction costs of financial intermediation for both short term and long term capital and especially for MSME;
- Assist SMSE to insert into national and global value chains in view of trade liberalisation;
- Capacity-building at the institutional and industry levels for SMEs;
- Initiate public private partnerships and improve the dialogue between them;
- Capacity building for public and private sector;
- Trade and investment promotion;
- Encourage backward and forward linkages between the EU and the countries of ACP regions and support Cluster and Networking Development;
- Assistance to comply with the requirements of the WTO TBT and SPS agreements.

TABLE 7.4. Estimated adjustment costs by region

Region	Fiscal Adjustment	Export Diversification	Employment Adjustment	Skills/Prod. Enhancement	Total Adjust. Costs
Central Africa	320	307	193	265	1085
West Africa	925	682	416	690	2713
East South Africa	775	702	375	630	2482
Southern Africa	340	261	217	255	1073
Caribbean	355	189	134	195	873
Pacific	210	175	82	175	642
Gross Total	2.925	2.316	1.417	2.210	8. 868[24]

Source: Adapted from Milner et al. (2005)

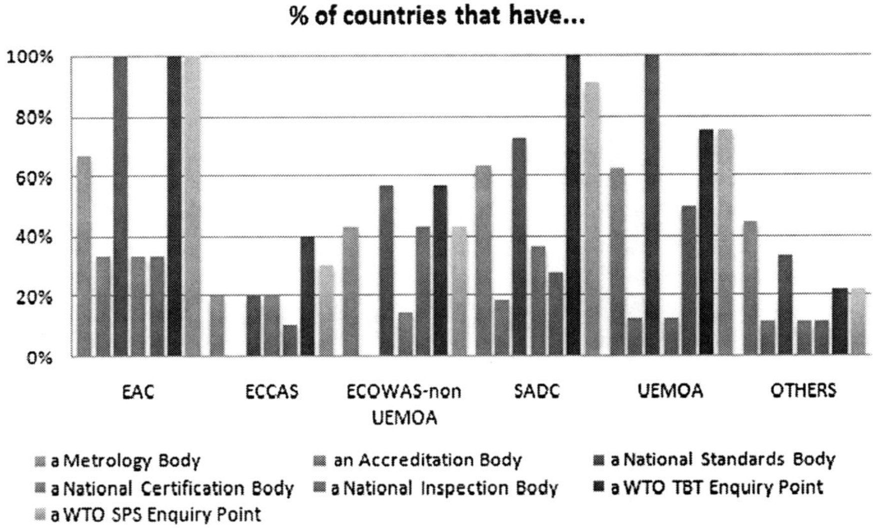

FIGURE 7.5. QI institutions in SSA (*Source:* UNIDO survey, UNIDO database and website info)

7.3. OPPORTUNITIES AND CHALLENGES OF EPAs IN AN INDUSTRIAL CONTEXT

7.3.1 Regional Integration

Article 35 (2) of the Cotonou Agreement states that "*Economic and trade cooperation shall build on regional integration initiatives of ACP States, bearing in mind that regional integration is a key instrument for the integration of ACP countries into the world economy.*"

The EU is striving to establish North-South-South arrangements. At sub-regional level, Customs Unions (CU) with a Common External Tariff (CET) are envisaged in most of the RECs.

In 2005, only 11 per cent of ACPs trade occurred within the region, while at the same time it was 66 per cent within the EU. In CEMAC and COMESA, two EPA negotiation regions, intra-regional trade is currently below 5 per cent (see Figure 7.4).

The Way Forward Addressing and solving political tensions and conflicts, especially in Sub-Saharan Africa, are the basic pre-requisites for the development of viable economic communities. Further regional integration in Africa can be a means for achieving industrial modernisation by encouraging intra-regional trade and securing economies of scale through facilitated market access. The regions should seize the opportunities arising from pooling resources together rather than developing incompatible regimes and policies. The strengthening of regional integration should contribute to enhancing productivity and competitiveness through cluster development methodologies, creating linkages with broader national and international value chains, promoting domestic and foreign investment and diffusing improved technologies.

The elimination of intra-regional trade barriers should provide the basis for consequent trade development and welfare enhancement both at national and regional levels. Harmonisation within the RECs will be necessary to determine more precisely the sensitive sectors to be protected and the sectors where further trade liberalisation will be possible within the EPA.

Sensitive Sectors Regional negotiating groups in the EPAs may consider putting those products on a sensitive list for the transitional period, if huge amounts of this product were previously imported from the EU or are expected to be. If these products were subject to high import tariff rates, they were consequently generating important tariff revenues. In some countries, the 20 per cent most profitable products cover more than 70 per cent of the revenues. Increasing imports in ACP countries, in absolute terms or in relation to the national production, could cause a serious damage to sensitive sectors both at country and regional levels. As a result of improved market access, some sub-sectors in ACP, contributing substantially to GDP, may enjoy particularly high growth rates in terms of EU imports, and consequently placed on the exclusion list as well.

Products from the agricultural sector are particularly sensitive since they contribute to income generation and national wealth. European agricultural products, which are sometimes highly subsidised, will directly compete to local products in labour-intensive sectors where the poorest population lives. For example, cotton, crops, oilseeds or vegetables may be considered as sensitive products. This issue is particularly critical since the agricultural sector remains, overall, the single largest contributor to employment in ACP states.

The manufacturing sector, which also employs a large number of people could be at risk as well. A further price decrease in the prices of products in that sector may harm producers.

Moreover nascent industries, with favourable prospects of sector and products development, should be protected. Increasing amounts of imports could threaten above all SMEs. It should be noted, however, that competition could lead to the largest efficiency gains, meaning that enterprises will be forced to be more productive and produce at lower costs.

Products should be classified in three categories taking into consideration their sensitiveness for each country: not facing competition between regions; competing but not strategic products; and, competing and strategic products.

The transitional period and the possibility to protect sensitive sectors for a certain period of time, should allow the development of intra-regional markets and subsequent to seize economies of scales necessary for international competitiveness.

7.3.2 ACPs Productive and Trade Capacity

Under the Cotonou agreement, ACP countries have preferential access to the EU market for nearly all goods. 97 per cent of their exports enter the EU duty free. The remaining three per cent result from 919 tariff lines in sensitive sectors that are not qualified for preferences under the Cotonou agreement (Hinkle and Newfarmer, 2005).

The EPAs would give ACP countries the opportunity to retain and further improve their current level of access to the EU. The EC announced to offer full market access[19] to ACP regions in EPAs negotiations, meaning that all remaining quota and tariff limitations will be removed immediately[20].

This is especially important for the 37 developing, but not least-developed ACP countries, since they don't fall under the Everything But Arms (EBA) initiative of the European Union. This initiative allows LDCs to export all products (except arms) duty free and quota free to the EU. Without the successful conclusion of EPAs, non-LDC countries could still export under the GSP, where the preferences vary depending on a product being categorised sensitive or non-sensitive. GSP is a voluntary scheme, with preferential access to the EU market applying to all developing countries. However, the GSP does by far not offer the same level of market access than under Cotonou. The average tariff is at 20 per cent so that in the case of West Africa more than €1 billion of trade would potentially be lost.

Nevertheless, EPAs will further improve market access since periodically reviewed conditions like in the EBA and GSP scheme could be subject of possible market access withdrawal. Not only it causes uncertainty for the local producers in the ACP countries, but also discourages foreign enterprises to invest in certain sectors. This, in turn, leads to a less diversified export structure. Also, the 919 tariff lines included sensitive sectors, such as agro-processed food, where many ACP countries could expand export activities. EPAs are therefore an opportunity to enhance market access and keep it at a constant level.

Anyway, the reason why ACP states still play a minor role in world trade, and did not reap benefits from Cotonou, cannot be found in the remaining existence of tariff barriers. The key hurdle are rather supply-side constraints that have urgently to be removed in order to ensure necessary quantity and quality supply and to strengthen export supply capacities for intra- and interregional markets. Standard compliance is an essential mean for this, in order to conform to international market requirements.

Seizing market access opportunities leads to an increase of economic growth what in turn not only increases government revenues but also has the potential to create productive employment opportunities and to improve household incomes.

The former Secretary General of the UN, Kofi Annan, claimed that "LDCs have neither the surplus of capacity of exportable products nor the production capacity to take immediate advantage of new trade opportunities"[21].

Supply-Side Constraints Micro, small and medium enterprises (MSME) are the main pillar in ACP countries, given that, overall, the majority of the population is working there and their value-added per employee in dollar terms is three to four times higher than in larger enterprises. However, they have a lack of productive capacities.

Various reasons hinder the establishment of internationally competitive enterprises: at the individual level these are a generally weak health system unable to reach people in rural areas, hunger and malnutrition and low education levels of the working age population. At the enterprise level an unfavourable financial situation with difficult access to loans, old technology with ill-equipped capital goods and inadequate managerial skills constitute the major constraints. Furthermore, landlocked or small-island states face special problems since they are remote from international markets and have a poor physical infrastructure.

The ACP countries, weak in terms of human resources, are relatively well endowed with low-skilled workers generating opportunities for labour-intensive methods. Better use of production factors, including financial, natural, human and physical capital, will help to expand production.

Enterprises have to produce at competitive costs by being more productive. Manufacturing capability in terms of price, quality and innovation has to be created. Therefore, the private sector in ACP countries needs to provide a conducive legal and regulatory framework, which supports private enterprises, especially MSMEs.

Furthermore, the export structure in ACP countries needs to be corrected over time, by seeking for new production possibilities. Especially in resource-intensive countries, it is essential to move away from unprocessed raw materials towards more value-added in export-processing industries. Sustained economic growth is associated with the capacity to diversify domestic production structure.

Countries must have marketable products for exportation. Competitive, safe, reliable and cost-effective products have to be offered to world markets, until now absent in most ACP enterprises. This can be achieved through efficient upgrading of the industrial productive capacity at both the institutional and enterprise levels.

In order to connect ACP countries successfully to non-domestic markets trade capacities have to be created and maintained. Without such capacities ACP countries will be unable to benefit from the new market opportunities arising from the EPAs. It is recommended to focus on sub-sectors with a high export potential. A weak financial system is reflected in the high transaction costs of financial intermediation that reflect the underlying risk associated with political and economic instability, weak legal systems and informational problems.

EPAs offer the right framework for addressing both supply-side rigidities and demand-side constraints. Adjustment measures should generate new activities and enhance the role of manufacturing in economic transformation. This includes UNIDO's Industrial Upgrading and Modernisation Programme that has the potential to strengthen national capacities and maintain the dynamic process of integration of exports, employment and industrial growth in the context of globalisation and ACP trade liberalisation, thus making an effective and coherent contribution to development. This approach is also in line with the African Productive Capacity Initiative (APCI), adopted by the New Partnership for Africa's Development (NEPAD) as its industrial component.

Quality Infrastructure, SPS and TBT Once the critical export quantity is achieved the problem of reaching a certain quality requested by international markets and the associated challenge of proving conformity with international standards and technical regulations arise.

This requires establishing efficient testing, calibration, certification and accreditation mechanisms that conform to the requirements of the SPS and TBT agreements and enjoy international recognition. This takes on extreme importance for ACP countries wanting to benefit from trade opportunities.

ACP countries have to prove compliance with quality standards and the adequate Quality Infrastructure (QI) has to be put in place. QI is defined as the combined institutional framework related to standardisation, quality management, metrology, testing, inspection, certification and accreditation. The overall objective of the QI is to improve the quality of products, processes and services and to effectively demonstrate through adequate conformity assessment procedures that these products, processes and services comply with requirements such as standards or technical regulations.

However, if the quality infrastructure is to effectively support exports, it has to be recognised internationally and link to the relevant international organisations.

UNIDO undertook a survey in 2006 showing that the level of development of the QI in Africa is quite low because very few countries have the totality of the adequate QI elements (see Figure 7.5). QI institution building should be done in conformity with international norms in order to remove technical barriers related to standards, technical regulations and conformity assessment procedures in trade. However, many African countries are not members of international organisations.

Many countries may have existing QI institutions, which, not for all intents and purposes, are operational. This, in turn, weakens all other conformity assessment areas such as certification or inspection.

Few certification bodies exist although certification worldwide has kept on increasing at a rapid pace. It includes standards for food hygiene and safety (ISO 22000), quality management (ISO 9001), environmental management (ISO 14001) and Social Accountability (SA 8000), as well as Corporate Social Responsibility (CSR). A pro-active approach is needed to keep pace with international requirements.

Thus, the following areas need special attention resulting from an insufficient QI infrastructure in ACP countries:

- Quality Promotion,
- Accreditation, metrology, standardisation, testing, inspection, certification, and
- Traceability.

Sanitary and Phytosanitary Standards (SPS) According to WTO, the aim of SPS is to protect human, animal or plant life or health within the territory of the Member from risks arising from pests, diseases, additives and contaminants[22]. The WTO SPS Agreement is the international legal framework on how to set and apply SPS measures in the international trading environment.

The hurdles that ACP countries face to conform with SPS requirements are too high and there is a concern that ACP countries will lose market access to the EU because of stricter SPS measures. Despite the considerable amount of ODA earmarked for SPS, ACP countries find themselves in a difficult situation since they are not able to make use of the instruments provided in the SPS Agreement to facilitate market access.

A careful treatment of SPS related issues in the EPA negotiating process has to be ensured, also because ACP countries will possibly seize newly arising export potential in the area of agro-food processing.

To reduce the trade distorting effects of SPS measures, the following issues should be addressed:

- Help ACP states to develop strategies in order to meet the most precluding SPS requirements standards;
- Establishment of inspection sides to check for signs of disease before shipment;

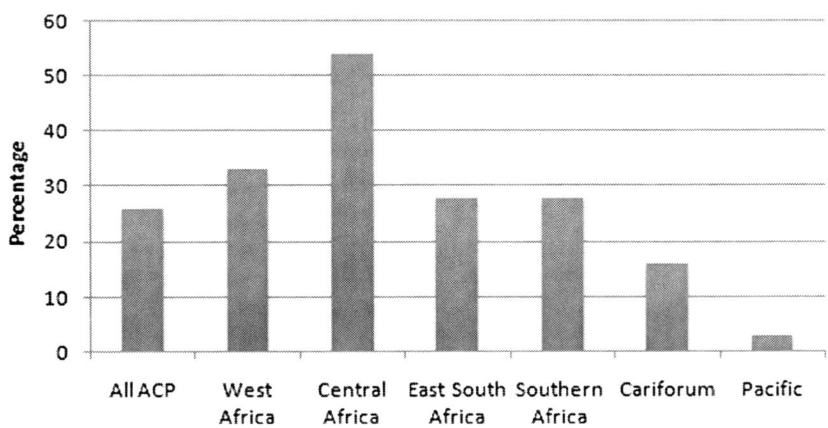

FIGURE 7.6. Imports from the EU (as % of Total Imports) (*source:* Comext (2005), except All ACP (2006 Comext) and East South Africa (2004 UNSD COMTRADE))

- Liaise ACP governments and EU importers for strengthening ACP producer's capacities to achieve SPS requirements;
- Development of food legislation, standards and regulations;
- Set up WTO SPS enquiry points and help deal with notifications and queries;
- Harmonisation of regional measures, regulatory frameworks and policies;
- Establishment and upgrading of regional and national testing authorities and accredited laboratories, etc.

Technical Barriers to Trade (TBT) The way a product is labeled or packaged before it is sold or the shape, size, design, functions and performance of a product are specified in the TBT Agreement of the WTO. They are deemed to be necessary, mainly by developed countries in order to protect consumer's interest and the environment.

Civil societies often argue that those notifications only serve to protect domestically produced goods and are therefore another obstacle for developing countries' market access.

However, in Article 47(1) of the Cotonou Agreement, the EU accepted to remove unnecessary TBTs in order to facilitate standardisation, certification and quality assurance compliance for ACP countries. The EPAs offer additional possibilities to address TBT constraints through financial and technical support to enhance capacity for:

- Development, implementation and monitoring of technical regulations;
- Establishing of WTO TBT enquiry points and help deal with notifications and queries;
- Helping ACP countries to comply with packaging, marking and labeling requirements;
- Harmonisation of national technical regulatory frameworks within RECs.

Rules of Origin Restrictive and burdensome rules of origin (RoO) are another factor that put burden on exporters in ACP states. They define the conditions that a product must fulfill in order to 'originate' from the exporting country, which is granted preferential access. The main motivation for applying RoO is to prevent "trade deflection".

The less protectionist RoO are, the less burdensome and costly it is for ACP producers to prove compliance. A simplification of RoO could facilitate market access, increase production, processing, quality and diversification of products. An opportunity for ACPs exists in the way that costs of production can be reduced if intermediary products can be imported. New production and export opportunities for value-added goods may arise for products that are not locally produced.

TABLE 7.5. European Union exports to ACP countries in 2006

European Union, Exports to ACP (South Africa excluded)	2006 (Mio €)	%	Share of total EU exports
Total	**34.033**	**100,0%**	**2,92%**
Primary Products	8.123	23,9%	5,33%
of which			
Agricultural products	4.471	13,1%	6,34%
Energy	3.082	9,1%	6,34%
Manufactured Products	25.195	74,0%	2,56%
of which			
Machinery	9.824	28,9%	2,92%
Transport equipment	5.343	15,7%	3,06%
Automotive prod.	2.623	7,7%	2,21%
Chemicals	3.676	10,8%	1,97%
Textiles and clothes	528	1,6%	1,44%

Source: EUROSTAT (COMEXT, Statistical regime 4)

7.3.3 Trade Liberalisation for EU Exports to ACP

With a successful negotiation, EPAs will lead step-by-step to export reciprocity between ACP and EU. The 27 European Member States will therefore be able to export substantially all merchandise at a zero-tariff rate to ACP countries. A free trade zone will be established between the EU and ACP after a transitional period whereas 80% and more of trade have to be liberalised. Tariffs, quotas and other restrictions will therefore be removed gradually. This concerns all types of products: primary, agricultural, and manufactured goods.

EU industrial exports to ACP and tariff dismantlement The EC argues that openness generates predictable and positive consequences for growth[23]. In view of the EPAs the EC states that the gradually opening of the ACP markets will improve ACPs competitiveness as a result of economies of scales, diversification of exports and increased efficiency, and at the same time reduce costs and therefore consumer prices. Economic growth for the regions and greater investment inflows will be the positive consequences.

Between 2002 and 2006, exports rose by more than €5 billion, with the highest increases in energy related products and machinery. Details of EU exports to ACP can be seen in Table 7.5 below.

The EU is the biggest import partner of ACP countries, constituting 24.7 per cent of all imports. For the African ACP countries the share is even a 31 per cent, and especially high in Central Africa (54%) and West Africa (33%) (see Figure 7.7 on the next page).

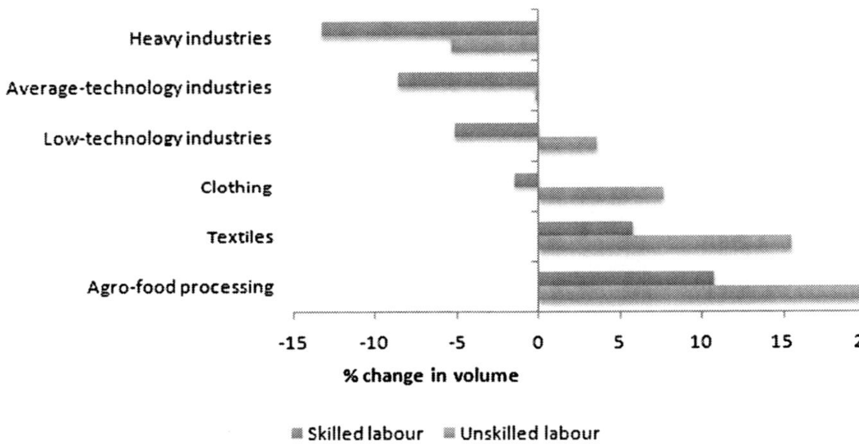

FIGURE 7.7. Demand for skilled and unskilled labour in SSA (volume terms) (*Source:* Data compiled from Kharingi et al. (2005))

Governments in ACP countries achieve a not negligible part of their revenue through income taxes, consumption taxes and tariffs on imports. Although the share decreased in the last years, in some countries, import duties still account for 50 per cent of total government revenue.

Given the high share of imports from the EU and ACP government's dependence on tax revenue, it is intuitively clear that an elimination of customs duties would have considerable negative effects on countries revenue. This will only happen after a transitional period of 12 years or more while an estimated 20 per cent can be left out of the liberalisation process.

Impact Assessment for Industries Industrialisation should play a key role in the EPAs if recent economic growth in ACP countries should continue to prevail in the next years and decades and therefore be sustainable. As outlined before, in most ACP countries the majority of people however live from agricultural and agro-processed products where high commodity price fluctuations change harvest earnings from year to year. If then also natural disasters wipe out crops of the entire year, like it happened in September 2007 in Western, Central and Eastern SSA due to flood, it is intuitively clear that a higher share of GDP should be obtained from the industry sector.

In a UNECA study Karingi et al. (2005) investigated the implications of EPAs on SSA industries by means of a general computable equilibrium model.

Tariffs in Sub-Saharan Africa (SSA) on imported industrial goods from the EU are considerably higher than vice versa, with the exception of agro-processed goods (for details see Table 7.6). As outlined before, industrial productive and trade capacities are already low in ACP countries. The schedule for abolishing customs duties is of importance for ACP countries implementing EPAs, as they set out the time span to proceed to internal industrial adjustments before liberalisation. Fast and imprudent industrial tariff liberalisation will lead to de-industrialisation and as a consequence lead to worsening of conditions of ACP livelihoods.

In the study a scenario has been simulated where all trade barriers between the SSA and EU are eliminated in both directions. This final stage of the EPAs, namely a free trade area, would only enter force after at least 12 years. They found evidence that there will be a serious risk of de-industrialisation, most eminent for heavy, low-technology and medium-tech industries if during the transitional period sectors at risk will not enjoy accompanying measure for becoming more competitive (Figure 7.7).

TABLE 7.6. Difference between EU and SSA industrial tariffs

	EU tariffs rates on SSA	SSA tariffs rates on EU	Difference
Agro-processed goods[24]	39,4	23,9	−15,5
Textiles	10,9	16,4	5,5
Medium tech industries	2,1	15,4	13,3
Heavy industries	1,4	15,8	14,4
Clothing	12,1	29,6	17,5
Low tech industries	2,6	23,5	20,9

Source: Karingi et al. (2005)

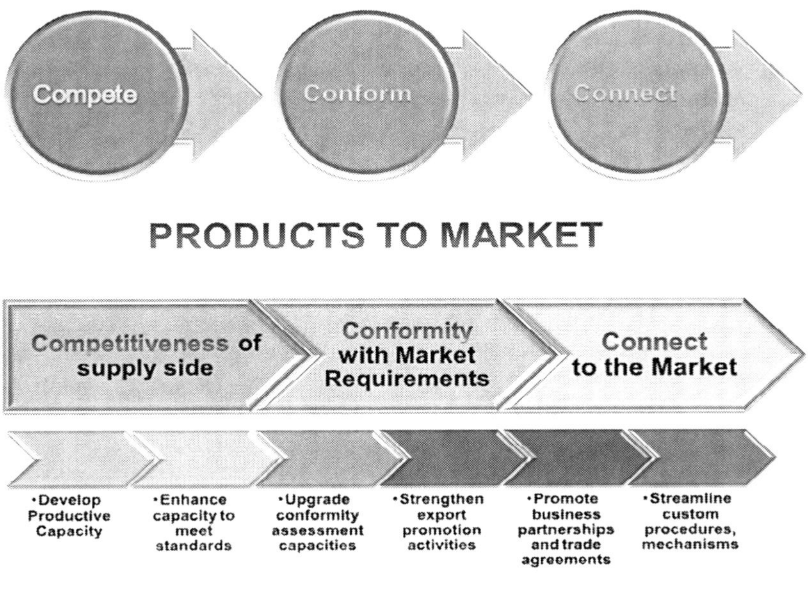

FIGURE 7.8. UNIDO's 3Cs Approach

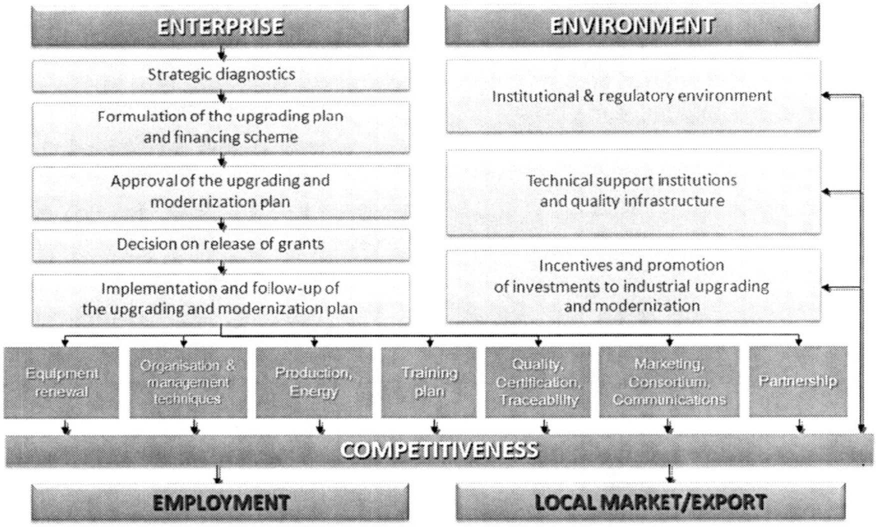

FIGURE 7.9. UNIDO Upgrading and Modernisation Approach

As a result of the present tariff difference, capital-intensive heavy industries and low-technology industries will witness a reduction in output, in average at around 10 per cent and 5 per cent respectively. These industrial sub-sectors would offer the potential for diversification in high value-added products but are already now the main EU exports products to ACP which are likely to increase.[25]

On the other side good perspectives exist for textiles, clothing and agro-food processing. These are labour-intensive manufactures where EU/ACP tariffs are not diverging too much. Growth rates in the order of 2–12 per cent are expected.

A similar picture is given if the endowment changes of skilled and unskilled labour are taken into account. Demand for mainly skilled labour in the heavy-, average-, and low-technology industries will contract as a result of outcompeted SSA enterprises. In manufacturing sectors clothing, textiles and agro-food processing mainly unskilled labour will be demanded. This results from an increase of output of these sectors and the specific demand of skills for these. Most ACP countries have high unemployment rates but are well endowed with low-skilled workers. These sectors will offer productive work for youth and stimulate the demand for high-educated university graduates.

Although these figures have to be handled with care due to data constraints and the limits of a computable general equilibrium models, they give a hint for the direction of industries in an EU/ACP free trade area.

ACP countries have the potential to build a strong industrial base but only if they have enough time, and, again, the right measures to build productive and trade capacities and diversify their industries. Therefore, upgrading and modernisation of industrial enterprises should be among the priority Aid for Trade measures addressed by the EU in order to create industries able to compete with imported European goods.

Other studies assessed the economic and welfare impact of EPAs focused on regions and countries within the EPA negotiation. Trade flows and government revenues changes in Sub-Saharan Africa have been examined.

7.3.4 Foreign Direct Investments (FDI)

According to UNCTAD's World Investment Report 2007, total Foreign Direct Investments (FDI) inflows in the world in 2006 rose by 38 per cent to $1.306 billion. However, the ACP region received only 1.56 per cent of total world outward flows in the year 2006.

The poor diversification of industries and concentration of FDI in primary sectors until now is not conducive for eradicating poverty. African FDI inflows are mainly resource seeking (oil and gas) and therefore delinked from the domestic economy. Profits obtained by the multinational corporations (MNC) are not reinvested; hence do not lead to economic development and increase of the manufacturing sector[26]. Few efficiency-seeking FDIs are coming to ACP countries. But there seems to emerge another group of investors in Africa. It is about SME whose investments are more beneficial and spread steadily between manufacturing and services. These enterprises have to be attracted in order to generate local productive employment and spillovers in manufacturing.

By fostering regional integration, EPAs are expected to pave the way for increased FDI. The EC points out that larger markets are key to attract further investments, since it means more potential customer[27]. The EC will finance infrastructure programmes under the 10th EDF, aiming at decreasing the Business Cost Factors and therefore render investment more attractive in ACP countries.

In the framework of EPAs several important issues related to enabling ACP countries to become more competitive in attracting and absorbing FDI have to be identified. Furthermore, strategy and policy advice on FDI and institutional capacity building are critical success factors for attracting investors.

7.4. UNIDO's RESPONSE TO EPAs

7.4.1 UNIDO's Productive and Trade Capacity Building Approach

Although its assistance is mainly provided at country level, UNIDO has developed over the years a regional approach through the establishment of Regional Programmes within its internal structure and with specific thematic programmes aimed to support ACP Regional Economic Communities (RECs).

The Long Term Strategic Vision adopted by UNIDO 11th General Conference in 2005 in Vienna has focused its work on three thematic areas where its contribution can be most effective:

(1) Poverty reduction through productive activities;
(2) Trade capacity-building;
(3) Energy and environment.

The following developments will be concentrated on the Trade Capacity Building area, which is the most relevant in the EPA context. However, the other thematic areas are also contributing to the regional programmes being developed by UNIDO.

Particularly established to support the preparation of developing countries and countries in transition for further trade liberalisation, the area of Trade Capacity Building is based on the 3Cs Approach: supply Competitiveness, demand Conformity, market Connectivity:

- **Developing *competitive* manufacturing capability** through upgrading of product and production quality, productivity enhancement, safety and cost-effectiveness improvement;
- **Developing and promoting *conformity* with market requirements** by establishing the necessary physical and institutional infrastructure to prove that products conform to the technical requirements laid out within the multilateral trading system with particular attention to the development of standards and conformity assessment infrastructures;
- **Enhancing *connectivity* to markets** through more effective participation in international trade negotiations and engagement in trade facilitation. In this area UNIDO cooperates with other organisations (UNCTAD, ITC).

Enhancing productive capacity is one of the most realistic and effective approach to ACPs sustainable participation in global production networks and segmented systems. Value-added needs to be generated at each stage of the segmented production process.

The overall objective is to transform comparative and competitive advantages into cumulative advantages for ACP.

UNIDO's aim is to promote export-oriented production in order to eradicate poverty and to achieve the MDGs. It focuses on increasing developing countries' manufactured exports and value-added. Technical assistance for the upgrading of their institutions, support infrastructures and competitive productive capacities are provided in order to: facilitate their integration into the world market, overcome both tariff and non-tariff barriers to trade and, ultimately, create a favourable trade environment.

7.4.2 UNIDO Approach in the EPAs

In order to build a consensus on the needs assessment and on the trade related technical assistance necessary to prepare ACP regions for further trade liberalisation, UNIDO has developed its programmes in partnership with the RECs. As a technical partner, UNIDO is not involved in the EPA negotiations but its first role is to provide technical assistance to its Member States. Both ACP RECs and the EC have been involved, since the very beginning of the programmes identification, in the elaboration of the Term of Reference of the various missions and were regularly briefed on the development of the programmes. This transparent mechanism has greatly helped to create a consensus on the analysis of the situation in ACP regions and on the adequate response to bring in terms of technical assistance.

In order to support its ACP Member States in a critical time, UNIDO has decided to invest important financial resources to formulate these regional support programmes. As its ACP Member States, UNIDO is convinced that regional integration and cooperation is essential for their progressive integration in the global economy.

Given its recognised expertise and successful experience in the area of industrial upgrading and modernisation as well as in the establishment of the Quality Infrastructure, UNIDO received requests from the 6 ACP RECs to assist in the formulation of regional programmes in the field of industrial upgrading and modernisation. The EU agreed on UNIDO's assistance for formulation and supported UNIDO to participate in the RPTF.

To date, 4 out of 6 sub-regional industrial and modernization programmes were validated by four African Regional Economic Communities. This was possible due to the work carried out in the field and in Vienna in close cooperation with RECs and Regional Preparatory Task Force (RPTF). Numerous meetings of the joint ACP/EU/UNIDO Task Force took place to discuss and amend the draft programmes taking into account the priority needs of the African RECs and their Member States.

7.4.3 Upgrading and Modernisation of Industries for ACPs

UNIDO has developed technical assistance programmes in the area of industrial restructuring, upgrading and modernisation to prepare and adapt industries and their environments to a changing global landscape, which is characterised by increasing competition. In the framework of EPAs, UNIDO's upgrading and modernisation of industries programmes aim at preparing enterprises for international competition and at improving productivity, quality, competitiveness and increasing export capacities. Furthermore they aim at enhancing and supporting competitiveness of productive sectors in terms of price, quality and innovation with focus on SMEs, integration in global trade and growth at the industrial and enterprise levels. The objective is to prepare industries for increased EU import competition in the framework of EPAs and the global market.

The programmes are a pragmatic and effective instrument for industrial and trade development. Also technical support institutions, such as Quality and Standardisation Institutes, will be taken into account and help them to comply with TBT and SPS requirements, among others.

The objectives of this programme are in line with: the Millennium Development Goals, the Doha Round and Doha Development Agenda and the African Productive Capacity Initiative adopted by NEPAD. Furthermore, the RECs are the owner of the programme and of its achievements and sustainability what is in line with the 2005 Paris Declaration on Aid Effectiveness.

The UNIDO programme of upgrading and modernisation is based on a multi-faceted approach and takes into consideration past programmes and lessons learnt.

It will be implemented through methods and tools adapted to the specific needs of ACP countries in the framework of EPAs with the following three integrated components:

Component 1: Support the upgrading and the competitiveness of industries and related services;

Component 2: Enhance capacities of quality infrastructure;

Component 3: Establishment/strengthening of the institutional and technical capacities and organs in charge of the restructuring and upgrading processes.

The outcomes of the three Components are:

Component 1: Support to upgrading and modernisation of small and medium manufacturing enterprises and related services in high-potential sectors and markets largely contributing to regional and national economies. Industries in ACP will be assisted through soft and hard upgrading activities in order to face regional and international competition and take advantage of the positive effects of trade liberalisation. Soft upgrading activities include inter alia: strategic diagnostic, formulation of upgrading programme, assistance for the implementation of priority actions related to products and the production system, quality, marketing, information management, finance, traceability, partnership. Hard upgrading activities include assistance for modernisation of equipment.

Component 2: Establishment/strengthening of regional and national system for accreditation, standardisation and quality infrastructure. Accreditation activities include: upgrading capacities of laboratories and setting up regional networks of specialised laboratories, harmonisation of analytical procedures for analysis and testing. Standardisation activities include: establishment of regional/national documentation centre for standards, harmonisation of standards and technical regulations, development of a framework law for consumer protection. Quality promotion activities include: annual quality award, assistance for certification ISO 9000 of industries. Regional and national capacity in the field of quality will be developed.

Component 3: Diagnostic study and upgrading programmes will be elaborated and implemented for technical support institutions to improve capacities of institutions to assist in the upgrading and development of enterprises mainly in the areas of: technology, investment, export, vocational training, etc. Centres of excellence to support enterprises, strengthen the interaction between research and industry, use of technology to develop innovation will be promoted.

UNIDO's industrial upgrading and modernisation approach Improvement of all competitiveness factors in the upgrading process is envisaged. For that purpose, an analysis at the macro, meso and micro level will be made at the initial stage. Several crucial factors for a successful outcome have to bear scrutiny.

The figure below gives an overview of UNIDO's Upgrading Approach.

The initial stage has four consequent steps: the pre-diagnostics, the strategic diagnostics, support to the formulation and implementation of the plan for upgrading and modernisation in close cooperation with the enterprise.

Upgrading technical support institutions Upgrading technical support institutions is a key prerequisite to improve marketing and productive capacities and competitiveness of industries.

The majority of companies in the ACP region suffer from considerable technology gaps mainly characterised by a lack of know-how on technical procedures and quality/safety management systems along with the absence of innovation initiatives in introducing new products and technologies. Moreover, ineffective exploitation of productive capacities and inadequacies in equipment maintenance lead to fabrication of defective goods or to the total suspension of production process.

Often overwhelmed by these technical and organisational shortages, the companies are limited in their capacities to optimise their technological process that determine their competitiveness and their place on the market. As a result, these companies are unable to produce goods that can response to the current market needs and meet international quality standards, sanitary and safety requirements.

One of the main reasons for this appears in the weak technical skills and qualification of the technical support institutions. Their technical capacities are insufficient to satisfy the real needs of industrial sector.

The current situation analysis shows that the national R&D and training institutions, the specialised technical centers and laboratories, the controlling, auditing and certification bodies do not possess the necessary capacities to support the companies, which, in consequence, have to seek the foreign expertise that brings to the companies additional costs and thus affects their competitiveness.

It is, therefore, crucial to address capacities of technical support institutions within implementation of upgrading and modernisation programme.

The following results are expected:

- Technical centers and training institutions strengthened or established in various industrial branches;
- Critical mass of national and/or regional expertise trained and operational;
- Regional network of institutions for technological support established and operational.

Box 7.1. The Tunisian national upgrading programme

- To date 2300 enterprises with a total upgrading investment of \$3.4 billion have benefited from this programme;
- Total upgrading grants: \$500 million;
- Increase in employment in beneficiary enterprises: 12%;
- Total industry exports increased from \$3 billion to \$11 billion;
- Exports for beneficiary enterprises increased by 18% per year; FDI increased from \$24 to \$327;
- 11% of enterprises that have started exporting did not so before the upgrading programme;
- 8 technical support centres have been upgraded in the sectors textile, leather and food and technical support institutions upgraded and modernised;
- 900 enterprises have been certified with ISO 9001;
- 20 laboratories accredited to ISO 17025;
- Capacity building in the area of upgrading has been provided for more than 500 experts.

Enhance capacities of quality infrastructure In the area of quality infrastructure, the following priority areas and the respective activities needed are overall needed in the ACP region.

Quality Promotion

- Training in the field of quality and quality management
- Establishment of legislation for quality control and consumer protection
- Undertake quality awareness and promotion campaigns at the enterprises and consumers levels

Metrology

- Acquire or upgrade metrology calibration and measurement equipment
- Training of the laboratory staff
- Set up metrology structures

Accreditation

- Establish internationally recognised accreditation structure
- Train the local human resources
- Acquire IAF and/or ILAC recognition/Membership

Standardisation

- Training of the standardisation structures' staff
- Creation of standardisation structures
- Raise awareness on the role and importance of standards

Analyses & Testing

- Acquire and/or upgrade laboratory equipment
- Training of the laboratories' staff
- Set-up analysis, testing and/or quality control centers
- Engage in inter-comparison activities with other laboratories

Certification

- Training of staff and human resources in general
- Acquire and/or upgrade equipment/instruments
- Set-up certification body

Inspection

- Training of staff and human resources
- Acquire and/or upgrade equipment
- Establish inspection legislation

Traceability

- Training in the field of traceability and/or about EU Traceability Regulations
- Acquire necessary equipment to undertake traceability activities
- Establish traceability legislation

Box 7.2. Major Milestones of the Restructuring and Upgrading Programme in Senegal

- Over 200 enterprises participate in the programme;
- \$44 million have been approved;
- To date, 32 enterprises benefited with a total sum of \$9,6 million;
- One technical support center has been upgraded;
- 45 enterprises received ISO 9001 certification;
- 160 experts received training on upgrading.

7.5. CONCLUSIONS

Time is an inherent problem concerning poverty in developing countries in general and ACP countries in particular. The deadline for signing Economic Partnership Agreements (EPAs) is approaching whereas the world economy is further gaining momentum. In fact, globalisation and intensifying world competition are challenging ACPs goal to reverse the trend of trade marginalisation, de-industrialisation and unemployment in their countries. At the same time, the EU is stipulating trade agreements with other regions in the world, leading to ACP preference erosion. If ACP countries do not manage to become more competitive in the near future they will not be able to sell their products in their own region and on international markets, but instead suffer from increased import competition.

By the end of 2007 there had to be a WTO-compatible trade agreement between the EU and all ACP regions although some countries do not yet see themselves in a position to sign full comprehensive EPAs. Time till then and the transitional period before complete tariff liberalisation should be utilised effectively and efficiently for the implementation of accompanying measures. Strengthening productive and trade capacities, establishing adequate quality infrastructure and upgrading technical support institutions are of utmost importance for integrating ACP into the world economy and ensure the timely achievement of the Millennium Development Goals (MDGs).

Concerning UNIDO's role in this context, its technical assistance in the thematic area of Trade Capacity Building is based on the "Aid for Trade" concept and on the 3Cs Approach (Compete, Conform, Connect). The Industrial Upgrading and Modernisation programmes aim at improving all factors of competitiveness by providing capacity building at the enterprise and institutional level.

The technical know-how and experience that UNIDO has acquired in programme conceptualisation and implementation in the field of building supply capacity and conformity infrastructure, together with the Organisation's extensive field presence, are valuable assets in the context of EPAs. UNIDO stands ready to contribute to the efforts of the 6 regional economic communities, the EU, civil society and bilateral development partners to make EPAs a success for ACP countries in general and for its least developed countries (LDCs) in particular.

NOTES

[1] Art. 1 of the Cotonou Agreement

[2] Art. 34 (1)-(3) of the Cotonou Agreement

[3] Eurostat, 2007 and World Bank, 2007

[4] With transitional periods for rice and sugar

[5] e.g. European Parliament, 2007 Report on EPAs, Committee on International Trade, Rapporteur: Robert Sturdy, 27.3.2007.

[6] South Africa is negotiating EPAs with SADC but already signed a FTA with the EU in 1999. Somalia and Timor-Leste, while being ACP countries are not negotiating EPAs.

[4] 12 new members since the beginning of EPA negotiations

[5] Including Somalia and Timor-Leste who are not negotiating/signing EPAs

[6] Barbandos, Seychelles, St. Kitt and Nevis, Bahamas, Tonga, Trinidad and Tobago, Antigua and Barbuda, Mauritius (with decreasing scores)

[7] 12 new members since the beginning of EPA negotiations

[8] Including Somalia and Timor-Leste who are not negotiating/signing EPAs

[9] Barbandos, Seychelles, St. Kitt and Nevis, Bahamas, Tonga, Trinidad and Tobago, Antigua and Barbuda, Mauritius (with decreasing scores)

[10] Source: Eurostat 2007

[11] IMF, Regional Economic Outlook: Sub Sahara Africa, April 2007

[12] EC, EPAs, Start of negotiations, A new approach in the relations between the EU and the ACP countries, 2002.

[13] OECD, WTO: Aid for Trade at a Glance 2007 (1st Global Review), 2007.

[14] EC, Communication from the Commission to the Council, the European Parliament, the European Economic and Social Committee and the Committee of the regions, Towards an EU Aid for Trade strategy – the Commission's contribution, 2007

[15] EC, DG EuropeAid, AIDCO Annual Report 2006

[16] Milner et al., "Some Simple Analytics of the Trade and Welfare Effects of Economic Partnership Agreements", Journal of African Economies, 14(3), 2005

[17] Only available for LDCs

[18] e.g. Draft EU – Eastern and Southern Africa EPA, July 2007

[19] Exceptions will remain for rice and sugar that will have a transitional period.

[20] EC, "EU offers full market access to Africa, Caribbean and Pacific regions in EPAs negotiations", 2007.

[21] Financial Times, 5 March 2001

[22] See "The WTO Agreement on the Application of Sanitary and Phytosanitary Measures".

[23] EC, Trading for Development - European Union/ Pacific Economic Partnership Agreement", 2007.

[24] Meat: cattle, sheep, goats, horse; Meat products nec, Vegetable oil and fats, Dairy products, Processed rice, Sugar, Food products nec, Beverages and tobacco products.

[25] UNCTAD, World Investment Report, 2007.

[26] UNCTAD, "Economic development in Africa: rethinking the role of FDI", 2005.

[27] EC, "EPAs: Means and Objectives", 2003.

REFERENCES

Hinkle, L. and R. Newfarmer (2005). Risks and Rewards of Regional Trading Arrangements in Africa: Economic Partnership Agreements (EPAs) Between the EU and SSA, World Bank.

IMF (2007). Regional Economic Outlook: Sub Sahara Africa, IMF, Washington.

Karingi et al. (2005). Economic and Welfare Impacts of the EU-Africa Economic Partnership Agreements. African Trade Policy Centre No. 10.

Oxfam (2006). Unequal Partners: How EU–ACP Economic Partnership Agreements (EPAs) could harm the development prospects of many of the world's poorest countries, Briefing Note.

Milner, C., O. Morrissey and A. McKay (2005). Some Simple Analytics of the Trade and Welfare Effects of Economic Partnership Agreements. *Journal of African Economies*, 14(3), 327–358.

UNCTAD. Handbook of Statistics (2006–2007), Geneva: UNCTAD.

PART II

AID FOR TRADE: REGIONAL PERSPECTIVES

CHAPTER 8

AID FOR TRADE AND PUBLIC-PRIVATE PARTNERSHIPS IN THE ASIAN AND PACIFIC REGION

8.1. INTRODUCTION

Aid for trade has attracted increased attention particularly since the Ministerial Declaration of the Sixth WTO Ministerial Conference of the World Trade Organisation[1] made explicit reference to the need for aid so that countries could increase their capacity to implement WTO agreements, as well as build their supply-side responses in order to facilitate trade and benefit from enhanced market access. This chapter sheds light on the background and rationale for AfT, making particular reference to the Asian and Pacific region (section 8.2). In particular, the paper highlights the fact that AfT is a multi-stakeholder and multi-actor initiative in which Governments and the private sector have roles to play. For that purpose, the paper explores various modalities for AfT by Governments and business, and in particular public-private partnerships (section 8.3). Related to this, the paper reviews public-private partnerships in trade facilitation initiatives in the region, including the possible convening of the Asia-Pacific Forum for Efficient Trade by the end of 2007. Section 8.4 discusses the role of ESCAP and other United Nations agencies in AfT.

8.2. AID FOR TRADE: BACKGROUND AND RELEVANCE TO THE ASIAN AND PACIFIC REGION

8.2.1 Background

An oft-quoted slogan, to the point of overuse, has been: "trade, not aid" as it has been universally recognised that trade leads to growth and development. Since the end of the Second World War, Official Development Assistance (ODA) has continued to increase in the post-war period. But by the mid-1980s, it became evident that the impact of ODA had remained limited while trade barriers remained high. This prompted many developing countries to call for better market access. At the same time, concerns were expressed with regard to the effectiveness of ODA, the absorptive capacity of developing countries, and the perception that often aid was "tied" to donor-driven agendas and conditionalities, including the structural adjustment programmes of the Bretton Woods institutions. ODA flows were further put into

Marc Proksch, Economic Affairs Officer, Trade Policy Section and Noordin Azhari, former Chief, Trade Efficiency and Facilitation Section, Trade and Investment Division, United Nations ESCAP, Bangkok.

P. De Lombaerde, L. Puri (eds.), *Aid for Trade: Global and Regional Perspectives*,
United Nations University Series on Regionalism 2,
DOI 10.1007/978-1-4020-9455-2_9,

perspective by the rapidly rising flow of private funds, in particular Foreign Direct Investment (FDI). All these led to a decline in ODA flows and at the same time market access increased as a result of the Uruguay round and proliferating regional trade agreements. Soon it became apparent that even with better market access, developing countries may not have the capacity to compete internationally apart from the traditional commodity trade from which they were urged to diversify. Hence, trade-related official development assistance, once again assumed more importance and the "Aid for trade" concept gained momentum.

Since 1995 and the establishment of WTO, the impetus of Trade-Related Technical Assistance (TRTA) has continued to increase. The link between trade and development was emphasised in the preamble to the Marrakesh Agreement Establishing the World Trade Organisation, while the Marrakesh Declaration of the General Agreement on Tariffs and Trade (GATT) recognised "the need for strengthening the capability of GATT and the WTO to provide increased technical assistance in their areas of competence, and in particular to substantially expand its provision to the least-developed countries". Thus, while the principal function of WTO is not development per se, the organisation has been traditionally engaged in providing technical assistance to developing countries to ensure their effective integration in the multilateral trading system and to help them to benefit from it.

The emphasis on TRTA for the least developed countries (LDCs) received further momentum with the establishment of the Integrated Framework for Trade-Related Technical Assistance to Least Developed Countries at the First Ministerial Conference of WTO, held in Singapore in 1996. The Integrated Framework represents the first attempt at providing TRTA in an organised, multi-agency, multi-donor coordinated programme. The participating agencies are the International Monetary Fund (IMF), International Trade Centre UNCTAD/WTO, the United Nations Conference on Trade and Development (UNCTAD), the United Nations Development Programme (UNDP), the World Bank and WTO. The objectives of the Integrated Framework are as follows:

(a) To "mainstream" (integrate) trade into national development plans, such as the poverty reduction strategy papers of LDCs;
(b) To assist in the coordinated delivery of trade-related technical assistance in response to needs identified by the LDCs. The Integrated Framework is built on the principles of country ownership and partnership.

The Integrated Framework served as a pilot that laid the foundations for AfT. It has encountered numerous problems, notably the lack of additional funding to implement trade-specific programmes, but it did pave the way for "more" coordinated delivery of TRTA at the global level. Since 2006, efforts have been taken to strengthen the Integrated Framework and enhance its implementation.

The need for TRTA has been emphasised in subsequent ministerial declarations of WTO. In particular, the Doha Declaration,[2] launching the latest round of multilateral trade negotiations – the Doha Development Agenda (DDA) – restates that "international trade can play a major role in the promotion of economic development and the alleviation of poverty" (para. 2). Paragraphs 38 to 41 specifically deal with TRTA and identify specific purposes for which TRTA should be delivered. In particular, "the delivery of WTO technical assistance shall be designed to assist developing and least developed countries and low-income countries in transition to adjust to WTO rules and disciplines, implement obligations and exercise the rights of membership, including drawing on the benefits of an open, rules-based multilateral trading system" (paragraph 38). Paragraph 39 recognises the need for coordination and underscores "the urgent necessity for the effective coordinated delivery of technical assistance with bilateral donors, in the OECD Development Assistance Committee and relevant international and regional intergovernmental institutions, within a coherent policy framework and timetable". Paragraph 43 endorses the Integrated Framework "as a viable model for LDCs' trade development" and urges "development partners to significantly increase contributions to the Integrated Framework Trust Fund and WTO extrabudgetary trust funds in favour of LDCs".

Paragraphs 48 to 51 of the Ministerial Declaration of the Sixth WTO Ministerial Conference in Hong Kong, China again reiterate the ministers' high priority to the effective implementation of the Integrated Framework and their endorsement of the Integrated Framework as a viable instrument for LDCs' trade development, building on its principles of country ownership and partnership. In particular, ministers highlight the importance of contributing to reducing their supply-side constraints. Paragraphs 52 to 54 deal with TRTA, re-emphasising the need for TRTA and specifically invite the WTO Director-General "to reinforce the partnerships and coordination with other agencies and regional bodies in the design and implementation of technical assistance programmes, so that all dimensions of trade-related capacity building are addressed, in a manner coherent with the programmes of other providers."

Paragraph 57 of the above-mentioned Ministerial Declaration for the first time makes reference to "aid for trade" and states that AfT "should aim to help developing countries, particularly LDCs, to build the supply-side capacity and trade-related infrastructure that they need to assist them to implement and benefit from WTO Agreements and more broadly to expand their trade". As such, WTO members for the first time explicitly recognised that AfT and multilateral trade negotiations are inextricably linked. They also emphasised that AfT should not be linked to "quid-pro-quo" bargaining and that more aid should not be a substitute for increased market access benefits for developing countries. The Declaration called for a task force to make recommendations on how to operationalise AfT. After its formation, the Task Force presented its report to the WTO General Council on 27 July 2006.[3]

8.2.2 Conceptualisation and Operationalisation of the Aid for Trade: An Ongoing Process

The recommendations of the Task Force are comprehensive. The report states that "the scope of Aid for Trade should be defined in a way that is both broad enough to reflect the diverse trade needs identified by countries, and clear enough to establish a border between Aid for Trade and other development assistance of which it is a part". It identifies the following categories of AfT:

(a) Trade policy and regulations, including training of trade officials, analysis of proposals and positions and their impact, support for national stakeholders to articulate commercial interest and identify trade-offs, dispute issues, institutional and technical support to facilitate implementation of trade agreements and to adapt to and comply with rules and standards;
(b) Trade development, including investment promotion, analysis and institutional support for trade in services, business support services and institutions, public-private sector networking, e-commerce, trade finance, trade promotion, market analysis and development;
(c) Trade-related infrastructure, including physical infrastructure;
(d) Building productive capacity;
(e) Trade-related adjustment, including supporting developing countries to put in place accompanying measures that assist them to benefit from liberalised trade;
(f) Other trade-related needs.

The Task Force also identifies challenges and gaps and makes recommendations concerning the operationalisation of AfT, at the multilateral, regional and national levels. It recommends, inter alia, that the "demand side" and use value-chain analysis as one valuable tool be strengthened in order to identify trade needs. It makes reference to and recommendations for an enhanced Integrated Framework, which implies that the Integrated Framework forms an integral part of AfT.

With regard to the regional-level implementation of AfT, the report recognises that "many countries require cross-border infrastructure and regional policy cooperation to trade more effectively. The ability to identify cross-border and regional needs should be strengthened at the country, regional and multilateral level". In particular, the Task Force recommends strengthening the following functions in relation to regional, subregional and cross-border issues: "diagnosis of needs; costing of projects; preparation of project proposals; and the coordination of donor response, including brokering and co-financing of needs that at present are difficult to finance through country-based processes, (e.g., cross-border infrastructure and policy-integration projects)". In this context, it is recommended the establishment of a regional aid for trade committee, "comprising subregional and regional organisations and financial institutions, to oversee the implementation of the subregional and regional dimensions of Aid for Trade, to report on needs, responses and impacts, and to oversee monitoring and evaluation".

For simplicity sake and without prejudging the final outcome of current international efforts to define and operationalise AfT, AfT could be defined as financial and technical assistance from multiple sources for the following purposes:

(a) Increase the capacity of Governments to effectively negotiate, conclude and implement multilateral, regional and bilateral trade agreements, including accession to WTO
(b) Increase the production capacity of businesses and their capacity to effectively compete internationally, i.e., to enable them to effectively utilise gains from increased market access as a result of concluded trade agreements.

Within these two broad dimensions of AfT, many other subdimensions can be identified the importance of which depends on individual country needs. It is understood that the needs are most urgent in LDCs and other less developed countries with specific development constraints, such as being landlocked, those in transition to a market economy, those of small size and island developing countries.

The trade negotiation dimension of AfT centres on human resources development and institutional capacity-building, including strengthening the legal framework for WTO compliance and enhanced capacity to negotiate and implement WTO agreements and commitments. With the adoption of bilateral trade agreements, some of which contain WTO-plus commitments, this dimension can be understood to include research and analysis on the design and contents of WTO-consistent agreements as well as capacity-building to negotiate and implement such agreements.

The second dimension of AfT, supply-side capacity-building, has assumed centre stage. This dimension not only addresses the need to raise the productive capacity of businesses but also to improve the environment in which businesses operate in order to enable them to compete effectively in international markets, such as through better institutional support and trade facilitation. The inclusion of infrastructure development, including transportation, energy and communications, is a debatable matter and no clear consensus seems to exist on this issue, in particular as it is difficult to determine to what extent infrastructure development for trade purposes can be distinguished from other purposes. However, the Doha Development Agenda Trade Capacity Building Database includes this category separately from TRTA and capacity-building (TRTA/CB). Similarly, the issue of inclusion of debt relief for AfT purposes is also a moot point.

A recent WTO note[4] on AfT identifies three principles for future work on AfT:

(a) AfT must be a complement to, not a substitute for, results from the Doha Development Agenda;
(b) AfT must not have to compete for existing ODA flows with other development and poverty reduction priorities (this was a problem for the Integrated Framework);

(c) The case for attracting AfT to implement WTO agreements and build trade-related capacity more broadly must have the commitment of trade, development and finance ministers in developed and developing countries, including LDCs, and the support of private business if it is to live up to its promise of catalysing their trade-related investment and production.

While there appears to be consensus that AfT is important, an emerging issue is to identify those recipient countries that should receive priority. In this context, Asia and the Pacific so far has not figured prominently. However, a cogent case can be made explaining why this region can and should not be ignored. This case is taken up in the next section.

8.2.3 Aid for Trade in Asia and the Pacific

8.2.3.1 Overview According to the database on ODA of OECD Development Assistance Committee (OECD/DAC), total TRTA/CB worldwide was $1.7 billion in 2005. Africa has emerged as the largest recipient of TRTA/CB, accounting for 36.4 per cent of total, while TRTA/CB for Asia and the Pacific[5] amounted to almost $331 million in 2005, or about 19.3 per cent of worldwide TRTA/CB (Table 8.1) and only about 1 per cent of total ODA to the region.[6] This share has declined from a high of 30.6 per cent of global TRTA/CB in 2002. Of the total amount for Asia and the Pacific, the Pacific has received a negligible share, while South Asia received the highest share. The amounts have declined since a peak of $473 million in 2002. However, it is difficult to assess the exact amount of AfT due to the different categorisation of ODA and reporting formats among donor agencies and countries. For instance, according to the 2006 joint WTO/OECD Report on Trade-Related Technical Assistance and Capacity Building, a total of $3.1 billion was committed to TRTA/CB worldwide, a significant increase over the $2 billion committed in 2001. These figures were obtained from the Doha Development Agenda Trade Capacity Building Database (TCBDB), an initiative of WTO and OECD.[7] This database is much wider in scope and covers all global and regional programmes, including those for tourism. According to this database, total TRTA/CB for Asia and the Pacific (defined as Far East, South and Central Asia, including Caucasian countries and Afghanistan, and Oceania but not including the Islamic Republic of Iran and Turkey) amounted to about $420 million or 14.6 per cent of the global amount in 2005.

TRTA/CB, as accounted for by the above-mentioned statistics, is divided into two categories: (a) trade policy and regulations, and (b) trade development.[8] Trade development has routinely captured the largest share in any given region. In 2005, the total volume of commitments world-wide for trade policy and regulation amounted to $906 million, while for trade development the amount was $2.17 billion. For Asia and the Pacific (excluding the Islamic Republic of Iran and Turkey), the amounts were $135 million and $285 million respectively. However, when the much wider definition

TABLE 8.1. Regional breakdown of trade-related technical assistance and capacity-building, 2001–2005 *(Thousands of constant 2004 United States dollars)*

Region/year	2001	2002	2003	2004	2005
East Asia[a]	42,066	79,329	82,528	42,184	20,322
South-East Asia[b]	94,257	191,561	159,766	172,026	128,943
South Asia[c]	70,965	126,610	140,792	131,252	130,191
Central Asia[d]	109,370	62,809	64,310	90,674	46,211
Pacific[e]	8,423	12,486	4,617	7,550	5,043
Asia and the Pacific	***325,081***	***472,795***	***452,013***	***443,686***	***330,710***
World	**1,251,054**	**1,541,128**	**1,761,283**	**1,826,577**	**1,716,968**

Source: Doha Development Agenda Trade Capacity-Building Database of WTO and OECD.
Notes: [a]China, Democratic People's Republic of Korea and Mongolia.
[b]ASEAN countries minus Brunei Darussalam and Singapore but plus Timor-Leste.
[c]Includes all members of the South Asian Association for Regional Cooperation (SAARC) plus Afghanistan and Iran (Islamic Republic of).
[d]Includes Caucasian countries of Armenia, Azerbaijan and Georgia.
[e]Cook Islands, Fiji, Kiribati, Marshall Islands, Micronesia (Federated States of), Nauru, Niue, Palau, Papua New Guinea, Solomon Islands, Samoa, Tonga, Tuvalu and Vanuatu.

of AfT recommended by the Task Force is used (consisting of trade policy and regulations, trade development, economic infrastructure, building productive capacity and trade-related adjustments), total global committed AfT in 2005 amounted to almost $26 billion or a combined share of 36.5 per cent of global sector allocable ODA.

Country-level data only; regional assistance is excluded.

It is expected, though, that AfT will continue to grow in absolute terms, especially if DDA is concluded. However, concerns have been raised not so much about the volume of AfT but rather about its effectiveness.[9]

8.2.3.2 Relevance of Aid for trade for the Asian and Pacific region How relevant is AfT for the Asian and Pacific region? It is recognised that the region is home to the world's fastest growing economies, in particular China and India, which have amassed huge foreign exchange reserves. Against this, one question arises: compared with other developing regions of the world, is there really a need for TRTA and AfT in this region?

To answer the question, the following facts should be considered:

(a) Asia and the Pacific is the largest and most diversified region in the world, both geographically and in terms of population. Two thirds of the world's poor live in the region. While it hosts some of the world's richest countries, it is also host to some of the world's poorest.
(b) In this context, it is worth noting that the Asian and Pacific region accounts for 14 of the total number of 50 LDCs in the world. These countries face severe capacity constraints, since they do not necessarily have access to finance from the capital-surplus countries. In addition, despite significant progress, the incidence

of poverty remains high in some of the high-growth countries as well. It is worth noting that LDCs have significantly increased their share in AfT since 1995.

(c) Despite the availability of capital surpluses in some countries, such surpluses are not necessarily easily available for investment purposes and as sources of finance for development. In particular, foreign reserves by themselves do not constitute investment capital and some countries actually experience huge capital outflows to developed countries (in particular the United States) as a result of the long-term willingness of some countries to swap high-return domestic investment with low-yielding (United States) government bonds and other securities.

(d) The Asian and Pacific region also accounts for a large number of economies not yet members of WTO; however, 12 of these have observer status and are in negotiations for full membership.

(e) While the region may have more advanced developing countries and developing countries with a greater potential than those in some other regions, it could be argued that such countries may merit increased AfT to ensure that they can take full advantage of the emerging opportunities for trade. In other words, countries which show promise for development should be actively supported to enable them to keep the development momentum. Not surprisingly, middle-income countries have seen their share of total AfT grow significantly to almost 40 per cent since 1995. There is also evidence that, of all regions, the impact of AfT has been the greatest in Asia.[10]

(f) While the region is awash with capital, demand for investment capital tends to outstrip supply. Already the needs for infrastructure financing far surpass the availability of available resources, both public and private.

For that reason, there is a strong argument in favour of delivering AfT to those most in need in the Asian and Pacific region, in particular, because the region is the world's most dynamic and as opportunities for economic growth and development, as well as for economic integration through trade and investment are abundant. It would be truly unfortunate if those opportunities are missed due to capacity constraints, and inadequate AfT to address those constraints. For instance, there are strong indications that sustained technical and financial assistance from multiple donors played a significant role in facilitating the accession to WTO of LDCs such as Cambodia and Nepal. While Africa has the Joint Integrated Technical Assistance Programme (JITAP), Asia and the Pacific does not have such a programme. A convincing argument could be made that a similar programme should be developed for Asia and the Pacific as well. Such a programme should have strong linkages with the Integrated Framework, and ESCAP could take a leading role in the coordination of such a programme.

Before delving further into the specific role of ESCAP in this process, section 8.3 below considers some modalities for public-private partnerships which could enhance the effectiveness with which AfT could be delivered in the ESCAP region.

8.3. PUBLIC-PRIVATE PARTNERSHIPS IN AID FOR TRADE

8.3.1 Concepts and Modalities

AfT is a multi-stakeholder and multi-actor initiative in which Governments and the private sector have roles to play. In particular, it is increasingly evident that Governments cannot do the job on their own while the role of private capital as a source of finance for development has increased considerably.

Although trade negotiations seek to create new trading opportunities, the gains are meaningless unless the business sector actually benefits as a result of enhanced trade. However, businesses often find that they lack the capacity to compete effectively at the global or even regional levels. While AfT also seeks to develop government policies that would strengthen supply-side capacities, a cogent case can be made for businesses to participate actively in this initiative. After all, businesses are the principal agents for trade and investment.

AfT delivered through public channels remains limited, whereas the needs are real and increasing with the expanding complexities in the global and regional trading environments in an era of unabated globalisation and regionalisation. As trade and investment are primarily business transactions, the argument can be made that businesses could also assume an increasing role in providing technical and financial assistance to Governments, as well as to other, smaller and weaker businesses.

With regard to the provision of business support to Governments, public-private partnerships could be conceived which would set and formulate the national agenda for international trade negotiations. The government and business sectors could also collaborate in the implementation of trade agreements and commitments, including joint initiatives to reform and strengthen the national legal and institutional framework for trade and make it WTO-consistent. This would require active and institutionalised government-business dialogue and cooperation. The absence of a serious and inclusive dialogue in some countries has made the acceptance of international trade obligations more difficult at the national level and, hence, their implementation and compliance with their terms. Businesses, including small and Medium-Sized Enterprises (SMEs), have an important stake in a successful negotiation outcome, and Governments need in-depth knowledge of the needs and requirements of business to be able to formulate the intended outcome and negotiation strategies. If Governments face capacity constraints on effectively conducting those negotiations, businesses could and even should help out; this applies in particular to the more experienced and resourceful ones through the provision of expertise and assistance, for example in compiling the needed background information.

Public-private partnerships are also useful in developing supply-side capacities. For instance, in the agro-food sector, Governments (which provide the incentive and enabling legal environment), farmers (who need access to markets) and businesses (which sell seeds and provide farmers with technical support) combine market potential, technical competence and managerial skills to build the supply-side capacity of the sector.[11]

With respect to the supply-side dimension of AfT, businesses could also provide assistance to other businesses. This argument is particularly relevant when one recognises that competition takes place more and more at the level of supply-chains rather than at the level of individual businesses. Moreover, these supply-chains have assumed regional and global dimensions. Within those supply-chains, the larger businesses, not only transnational corporations but also large domestic enterprises, could benefit themselves by providing assistance to their smaller suppliers, such as the previously mentioned example of an agro-food business providing technical support to farmers. This could be done in the form of capacity-building to comply with prevailing standards in a given industry, as well as by providing financial support and training to facilitate the adoption by smaller businesses of the latest production technologies, management techniques and practices, and quality certificates among others. The supply-chain, after all, is only as strong as its weakest link. Businesses could help each other to optimise the strength of the whole supply-chain. This is indeed already taking place in some supply-chains, but could be more actively promoted. However, businesses could also help Governments and aid agencies analyse value chains to determine the need for AfT.

In considering modalities for business involvement in AfT one has to realise that increasingly the United Nations and official international organisations are assigned the role of the Government by donors in AfT programmes. Thus, while the modalities described below are based on business-government relationships, in some of them the Government's role can be complemented and/or replaced by the United Nations or international organisations. In general, there are two possibilities:

(a) Business to Government: businesses could provide financial and technical assistance to Governments in order to strengthen the latter's capacity to negotiate, conclude and implement international trade agreements and facilitate accession to WTO. They can also fund public projects implemented for this purpose by multilateral aid organisations, such as the United Nations agencies. Businesses could also help Governments (and United Nations agencies) to provide assistance to other businesses in order to develop their trade capacity either through bilateral aid or through multilateral organisations, such as relevant United Nations agencies or through the provision of direct technical expertise (e.g., by participating in technical assistance activities as resource persons/consultants at own cost).

(b) Business-to-business assistance: businesses fund and implement their own assistance within their own supply-chain or industry cluster. Such assistance already takes place. More often than not, it is transnational corporations which provide financial and technical assistance (including training) to their overseas suppliers, which are often SMEs in developing countries. Often, such assistance is provided to enable the suppliers to comply with the standards they set as buyers. Governments, in turn, can help by forging such linkages and strengthening international supply-chains which integrate their SMEs in the global economy.

Businesses will be naturally indisposed to provide funding for public projects without a clear financial return. After all, Governments are supposed to implement such projects on the basis of the revenue obtained, which to a large extent comes from taxation on businesses and their transactions, i.e., trade and investment. Often, taxation levels and the modalities for levying taxes are cited as barriers to business and investment. Business will therefore be understandably wary of providing additional funds to public agencies. However, such funding would not be without return. If Governments simply have no capacity to negotiate international trade agreements, the results may have adverse effects on business. Businesses therefore have a stake in being closely involved in the negotiation process. At the minimum they should provide in-kind assistance by maintaining a close dialogue with the Government and participate in consultations regarding negotiation needs. More likely though, and already increasingly common, is business support to other businesses, often in the same supply-chain, which is to everybody's benefit. Such support does not have to be financial (unless a return on investment is expected) but can consist of providing technical expertise to ensure that the whole supply-chain functions at optimum efficiency and delivers a final top quality product or service. Finally, business support does not have to be financial but can consist of technical expertise provided free of charge to Governments and aid agencies. For instance, members of the ESCAP Business Advisory Council have participated in ESCAP-organised forums at their own cost.

Businesses of course already indirectly provide financial support for AfT. As recognised by the International Conference on Financing for Development, held in Monterrey, Mexico, from 18 to 22 March 2002, trade is a major source of financing for development, including trade development. As businesses are the primary agents of trade, obviously their activities provide revenue for development purposes. The exact linkages other than through taxes of business revenue to financing for development need further research and elaboration to more closely identify modalities for business involvement in AfT.

Although consistent calls are made for an increase in global ODA levels, including ODA for trade purposes, ODA remains limited and is subject to periodic declines. In fact, where ODA has increased it is often linked to foreign policy and security objectives, e.g., eradication of terrorism or humanitarian objectives such as disaster relief rather than for purposes such as AfT. Therefore, the public capability to provide sufficient funding for AfT will remain limited. As a result, and in their own interest, the role of business in AfT is expected to increase.

One of the areas which have been generally recognised as providing significant contribution to trade, growth and development is trade facilitation. It is now accepted as fact that, in countries where trade facilitation is not developed and functional, trade cannot flourish, and without trade, economic growth lags behind. Therefore, it is relevant to explore further how public-private partnerships could be harnessed to increase the effectiveness of AfT in the area of trade facilitation.

8.3.2 Public-private Partnerships in Trade Facilitation

8.3.2.1 Overview During the 1980s and 1990s, many small developing countries undertook comprehensive programmes of policy reform, structural adjustment and reducing trade barriers. Although open trade is essential for success in trade, they are far from sufficient. Countries also need infrastructure and institutions that will enable them to take advantage of export opportunities and compete in the global trading system. However, since its inclusion in DDA, more attention and recognition has been given to the importance of trade facilitation as a tool for developing countries to participate in, and benefit more from, global trade. Moreover, trade facilitation is one very important work area of AfT.

Trade facilitation involves many customs and other trade procedures that are required to move goods and services across national borders. In many developing countries often complex customs, inspection and logistic requirements deter the use of the most cost-effective trade corridors. Inadequate trade facilitation systems create an "efficiency penalty". According to the World Bank, antiquated trade administration in poor countries, combined with the failure to adopt information technology in support of trade facilitation, accounts for a 7 per cent loss in the value of goods shipped. Facilitating trade requires simplifying administrative and commercial procedures for the movement of goods and services, investing in supporting infrastructure, and removing other barriers to trade flows. Both the public and the private sectors need to act to facilitate trade. Governments need to examine carefully their trade policies and procedures to ensure that these do not unnecessarily restrict trade. They must also commit to reducing delays and eliminating environments that foster illegal payments associated with implementing the policies and procedures that are in place. The private sector needs to lobby for reform and monitor the implementation of such reform.

In trade facilitation, collaboration among stakeholders, both private and public, enables the identification of critical trade and transport impediments and in defining alternative ways to achieve similar objectives at lower costs. Public-private partnerships can provide a forum where stakeholders can develop strategies to remove barriers to efficient trade. By involving private parties with competencies in trade and with vested interests in the success of their businesses, public-private partnerships can develop the most efficient and properly regulated supply chains.

The United Nations Centre for Trade Facilitation and Electronic Business put forth a recommendation specifically addressing the purpose, methods of creation, and operating structures of public-private partnerships under "Recommendation No. 4, National Trade Facilitation Bodies".[12] The recommendation emphasises the importance of these partnerships in order to among others:

(a) Identify issues affecting the cost and efficiency of a country's international trade;
(b) Develop measures to reduce the cost and improve the efficiency of international trade;
(c) Assist in the implementation of those measures;
(d) Provide a national focal point for the collection and dissemination of information on best practices in international trade facilitation; and
(e) Participate in international efforts to improve trade facilitation.

To be successful, potential parties to any partnership must overcome their traditional beliefs about Government and business intentions. For example, the trade community frequently hesitates to meet with Government representatives and express its thoughts and concerns for fear of possible retribution through government action. The Government often shares this reluctance for fear that the trader might somehow find a technicality in regulations that enables non-compliance with the rules. An environment in which trust exists between both parties provides a precursor to a good public-private partnership.

8.3.2.2 Public-private partnerships in trade facilitation in Asia and the Pacific As discussed previously, AfT often uses channels of public-private partnerships and this is also applicable for the trade facilitation area. Institutionally, public-private partnerships could be promoted at three levels, namely the national, subregional and regional levels. The functions of institutions at the subregional and regional levels are identical except for broader membership coverage.

At the national level, national trade facilitation bodies can represent their countries in international and regional initiatives aimed at facilitating trade. Many public-private partnerships created to improve the efficiency of the supply chain are formed by the government entity with the greatest interest in modernising the existing process. Smaller, growing economies can benefit greatly from the use of public-private partnerships in trade facilitation because those partnerships provide inputs for aspects of the economy that governmental ones would address in larger economies.

Given the multidisciplinary nature of trade facilitation and the involvement of a large number of stakeholders from the public and private sectors, the benefits of trade facilitation can be maximised, if it is undertaken based on public-private partnerships. Effective cooperation and coordination among all stakeholders is a fundamental prerequisite for the success of trade facilitation. A partnership between the public and private sectors could help to better define the real needs, improve transparency and information flows and encourage viable and sustainable solutions. Several members of ESCAP, including Azerbaijan, Armenia, Mongolia, Kyrgyzstan, Pakistan and Sri Lanka, have already established or are in the process of establishing national trade facilitation committees or other forms of a national trade facilitation coordination mechanism.

Public-private partnerships at the national level however should not be confined to the establishment of national trade facilitation bodies alone but may also extend to subsets of trade facilitation measures such as the operation of single-window systems. The Singapore TradeNet Systems, Dagang Net of Malaysia and Tradelink Electronic Commerce Ltd. of Hong Kong, China are prime examples of successful public-private partnerships in trade facilitation. There are other areas of trade facilitation where public-private partnerships can operate successfully. These areas should be further explored.

There are also many active subregional groupings such as SAARC, ASEAN and the Pacific Islands Forum. Since these groupings are all the creation of their own members, it would be useful if the various national trade facilitation committees from within such subregional groupings could be organised or grouped into a subregional trade facilitation forum, as they face many similar problems. As and when a region-wide forum is established, such groupings could then be invited as a member of that forum.

At all levels, trade facilitation networks that build on public-private partnerships could articulate appropriate recommendations for strategies and action plans for the implementation of relevant international and regional conventions, standards and tools. They could also develop regional knowledge in order to improve the trade facilitation environment of countries. Further, they could provide the stakeholders with a platform for coordination and harmonisation of trade facilitation measures among countries and subregions. For example, for those trade facilitation measures that employ modern information and communication technology solutions, such as automated customs procedures, electronic single-window facilities and paperless trade environments, ensuring their interoperability at the regional level would be crucial. In such contexts, such a regional network could play a vital role.

Several intergovernmental meetings have sought to establish an effective regional mechanism to encourage more collaborative efforts to improve the implementation of international and regional conventions, standards and best practices, as well as to harmonise trade facilitation policies and measures in the Asian and Pacific region. Initially, the establishment of such a regional collaboration platform was proposed by the participants of the UNDP-ESCAP Regional Consultation on Issues in Trade Facilitation and Human Development, held in Bangkok on 16 and 17 August 2005, the International Conference on Strengthening Regional Cooperation for Managing Globalisation held in Moscow from 28 to 30 September 2005, as well as the Commission at its sixty-third session in Almaty, Kazakhstan in May 2007. A similar request was made by some members of ESCAP at the second session of the Committee on Managing Globalisation which was held in Bangkok from 12 to 14 October 2005.

Following these requests, ESCAP has been promoting the establishment of a regional network of subregional and national trade facilitation committees and other coordination mechanisms in the Asian and Pacific region, based on public-private partnerships. Its mission is to contribute to the effective implementation of trade facilitation in a coordinated and harmonised manner, and thus to increase the level of trade and competitiveness of the countries in the region. This initiative, the Asia-Pacific forum for efficient trade, would be aimed at championing efforts and initiatives for boosting trade and competitiveness through the streamlining of trade and border-crossing operations, building capacity and enhancing cooperation in Asia and the Pacific.

AFET could provide countries in the ESCAP region with a regional platform and voice in defining trade-facilitation strategies and action plans at the regional and sub-regional levels as well as promoting the implementation of international and regional trade facilitation and border-crossing conventions, standards, best practices and other tools. It could also serve to showcase and exchange best practices and lessons learned in implementing trade facilitation measures and monitoring and evaluating the impact of these measures at the national level.

The above-mentioned aspirations could be achieved in the region through a network of government agencies, such as ministries of trade and finance, customs authorities and other organisations, including national trade-facilitation committees, chambers of commerce, single-window operators that are willing to commit their own resources to deliver products in order to cater to the trade-facilitation needs and priorities of policymakers and practitioners. The organisation of regular AFET events would provide opportunities for representatives of Governments, business, civil society and academia to discuss current problems and plan future directions.

8.4. THE ROLE OF ESCAP AND OTHER UNITED NATIONS AGENCIES

Various United Nations agencies, including ESCAP and the other members of the United Nations Trade Cluster, UNCTAD, the other regional commissions, UNDP, the United Nations Environment Programme, among others, have been mobilised to sustain the momentum towards the operationalisation and implementation of the AfT initiative. The United Nations system is supportive of the AfT initiative and its developmental goals, which include helping to achieve sustainable development and poverty reduction. Such commitments are contained in, inter alia, the United Nations Millennium Declaration[13] and the 2005 World Summit Outcome.[14] To this end, to be effective, the United Nations system needs to promote a more coherent approach that could result in coordinated implementation of the AfT initiative. Currently, collective efforts are being undertaken to formulate a coordinated United Nations approach to AfT within the context of the United Nations Trade Cluster.[15] The important role of regional commissions has been recognised in respect of the "experience and significant analytical capacity on trade matters in the regional commissions, including, in particular, in the context of regional and subregional trade agreements. The Commissions could thus contribute effectively to the Aid for Trade initiative, especially in relation to regional reviews and their follow-up."[16]

As one of the United Nations regional commissions, ESCAP has long maintained a strong trade and investment programme in recognition of the need for regional-level TRTA. This programme has addressed both the market access and supply-side capacity-building aspects of AfT. The mandate of ESCAP focuses on promoting regional economic cooperation. Because trade and investment are by definition cross-border economic phenomena, trade and investment have assumed a central role in the activities of ESCAP, in recent years within the theme group of managing globalisation. In parallel, the Task Force has clearly identified a need for a regional dimension to AfT. In this context, the following considerations are pertinent:

(a) The needs for AfT are enormous, especially in the area of training and skills development. No agency by itself has the resources or capability to address all those needs.
(b) At the same time, it is recognised that effective AfT requires a multi-disciplinary approach which few agencies possess. ESCAP's strength is exactly the availability of multi-disciplinary expertise.
(c) UNCTAD, as the principal United Nations agency in area of trade and development is the global forum for the integrated treatment of trade and development and the related issues of finance, investment, technology and sustainable development. WTO, on the other hand, is the platform for negotiating multilateral trade rules, monitoring their implementation and handling trade disputes. WTO is not a donor agency and though it provides technical assistance it is supposed to find global and regional partners for TRTA. The International Trade Centre handles strategic and operational aspects of trade development, focusing on exports and, unlike UNCTAD and WTO, it deals directly with the business sector. None of these global trade-related institutions has a regional office.
(d) ESCAP is the only United Nations agency in Asia and the Pacific addressing comprehensive trade, investment and economic integration issues with a regional focus. It addresses multilateral trade issues through concrete training in cooperation with WTO and addresses the proliferating bilateral and regional trade agreements in the region by tracking and rating them in an attempt to forge their consolidation and integration as building blocks of the multilateral trading system. In addition, ESCAP provides regional-level support for supply-side capacity-building in areas such as investment promotion, enterprise development and trade facilitation.
(e) In recent years, a number of regional and subregional agencies started trade-related programmes (e.g., Asian Development Bank, UNDP). In view of its long experience and accumulated expertise as well as analytical capacity in this area, ESCAP offers effective support to such agencies.
(f) ESCAP therefore is complementary to the other agencies rather than duplicative, and it stands to increase the impact of its activities through concrete partnerships. Such partnerships do indeed exist, such as the WTO/ESCAP Technical Assistance Programme. There is scope to expand further such partnerships as recommended by the Task Force.

Whatever its definition, AfT needs to be provided on multilateral, regional and national bases as the Task Force recommends. While WTO, UNCTAD and ITC are suited to provide such assistance at the global level, the regional commissions are ideally placed to fill the regional dimension, in particular as each Commission has a strong and long-established and well-recognised trade programme, as previously mentioned. At the national level, UNDP could take the lead. The Task Force makes a strong case for donor and agency coordination and effective monitoring and evaluation of AfT in all its dimensions. The role of the regional commissions will be further elaborated in the joint efforts of the United Nations agencies to operationalise the concept of AfT. In these efforts, due reference should be made to the recommendations of the Task Force.

In past years, ESCAP has launched various cooperative programmes with global organisations such as UNCTAD, WTO and ITC; subregional partners such as ASEAN, SAARC, the Economic Cooperation Organisation, the Eurasian Economic Community and Shanghai Cooperation Organisation; as well as with national institutes in member countries. Such cooperation constitutes an excellent example of the division of labour and joint efforts to produce greater impacts in servicing member countries.

NOTES

1 Held in Hong Kong, China, from 13 to 18 December 2005. See WTO document WT/MIN/(05)/DEC.

2 The Doha Declaration on Trade-related Aspects of Intellectual Property Rights Agreement and Public Health was adopted by the Fourth Ministerial Conference of WTO, which was held in Doha from 9 to 14 November 2001. See A/C.2/56/7, annex.

3 See *Recommendations of the Task Force on Aid for Trade,* WTO, WT/AFT/1 of 27 July 2006.

4 See "Aid for Trade: WTO Work Programme on Aid-for-trade", Background Note (WT/AFT/W/26), 29 May 2007.

5 Unless defined differently, the term "Asia and the Pacific" refers to the area covered by the ESCAP members and associate members <www.unescap.org/about/member.asp>.

6 The data from this database differ somewhat from those in the TCBDB and do not include cross-country and regional level assistance. However, the TCBDB does not provide country-level data and defines Asia as including West Asia and Middle East, making it difficult to disaggregate the data for ESCAP developing countries only.

7 <http://tcbdb.wto.org>.

8 Trade policy and regulations covers support to aid recipients' effective participation in multilateral trade negotiations, analysis and implementation of multilateral trade agreements, trade policy mainstreaming and technical standards, trade facilitation including tariff structures and customs regimes, support to regional trade arrangements and human resources development in trade. Trade development covers business development and activities aimed at improving the business climate, access to trade finance, and trade promotion in the productive sectors (agriculture, forestry, fishing, industry, mining, tourism, services), including at the institutional and enterprise level; <http://tcbdb.wto.org>.

9 See for instance, *Aid for Trade: Making it Effective*, OECD, Paris, 2006 <www.oecd.org/dataoecd/27/14/37198197.pdf>.

10 See, for instance, the Rapporteur's Report of the OECD Policy Dialogue with Non-Members on Aid for Trade: From Policy to Practice, Doha, 6–7 November 2006, para 57, <www.oecd.org/dataoecd/24/55/37819542.pdf>.

11 This point was made at the OECD Policy Dialogue with Non-Members on Aid for Trade: From Policy to Practice, which was held in Doha, on 6 and 7 November 2006.

[12] Adopted by the Working Party on Facilitation of International Trade Procedures, Geneva, September 1974. TRADE/WP.4/INF.33; TD/B/ASTF/INF.33 [Edition 96.1], and ECE/TRADE/242 of March 1999.

[13] See General Assembly resolution 55/2 of 8 September 2000.

[14] See General Assembly resolution 60/1 of 16 September 2005.

[15] A concept paper on this issue has been developed by the Cluster, and a joint event involving UNCTAD and the regional commissions has been organised in parallel with the twelfth session of UNCTAD, which has been held in Accra, from 20 to 25 April in 2008.

[16] See Summary of Conclusions of the United Nations System Chief Executives Board for Coordination on its first session, held in Geneva on 20 April 2007, CEB/2007/2-CEB (Rev. 24 May), para 38, p. 10.

ESCWA

CHAPTER 9

PROMOTING AID FOR TRADE IN WESTERN ASIA

9.1. INTRODUCTION

Aid for Trade (AfT) has gained considerable importance and has become a major item on the international trade agenda and development discourse. This is to a large extent in response to concerns expressed by developing countries and economies in transition with regard to their capacities to implement WTO Agreements and undertake necessary adjustments to increase net development gains from trade liberalization and emerging trade opportunities. The AfT initiative gained in relevance during the course of the Doha round of multilateral trade negotiations as developing countries faced increased adjustment and competitiveness challenges and difficulties in implementing existing and future World Trade Organization (WTO) Agreements, as well as an inherent lack of supply capacity and trade-related infrastructure deficits and deficiencies. Realization of the net development gains from increased trade also suffered setbacks due to the lack of mutually supportive trade, environment and development policies at the national level and limited capacity to integrate trade policy and practices into sustainable development dimensions. Subsequently, the Sixth WTO Ministerial Declaration endorsed the concept of AfT and appointed a Task Force to operationalize the concept. It also mandated the WTO to look into financing for the initiative[1].

The WTO Task Force issued a report with recommendations that was endorsed by WTO members[2]. Various initiatives by United Nations agencies, including United Nations Conference on Trade and Development (UNCTAD) and United Nation regional commissions, have been implemented to sustain the momentum towards the operationalization and implementation of the AfT initiative. The Economic and Social Commission for Western Asia (ESCWA), as one of the UN regional commissions, is also active in the implementation of the AfT initiative and its developmental goals of helping to achieve sustainable development and poverty reduction, as committed in *inter alia*, the Millennium Declaration and the 2005 World Summit Outcome. To this end, to be more effective, the United Nations System needs to promote a more coherent approach in order to achieve the coordinated implementation of the AfT initiative.

In this regard, the objective of this chapter is to offer a coherent United Nations system-wide perspective and a comprehensive contribution to the international policy debate on how to practically operationalize the AfT initiative and ensure that it makes an effective contribution to the development dimension of ongoing WTO negotiations. Regional and national perspectives of the ESCWA region, as well as

P. De Lombaerde, L. Puri (eds.), *Aid for Trade: Global and Regional Perspectives*,
United Nations University Series on Regionalism 2,
DOI 10.1007/978-1-4020-9455-2_10,

global perspectives, on the challenges and opportunities in requesting and supplying AfT, which is consistent with the recommendations of the WTO Task Force. The chapter will also attempt to provide policy recommendations on how the AfT initiative can assist ESCWA member countries in promoting trade as an engine for growth and further economic development.

First, the importance of development assistance and AfT in the ESCWA region will be briefly presented in order to give an overview of the utilization of aid in the region. This will be followed by a discussion on how aid is used for trade-related purposes in the region and particularly existing challenges, opportunities and supply-side constraints in the ESCWA region with regard to AfT. The chapter will conclude with identifying the role of ESCWA in promoting AfT in the region.

This chapter will not discuss the global AfT initiative and how it has evolved in the past, as these topics are covered in other contributions within this publication. Rather, it is an attempt to give an overview of AfT in the context of the ESCWA region as a whole and will therefore not focus on specific country cases. In light of the limited data available on the region concerning AfT, further research is needed to analyze the depth and breadth of AfT effectiveness in the region and its implications.

9.2. TRADE, DEVELOPMENT AND THE ESCWA REGION

The ESCWA member countries[3] have been gradually emphasizing the importance of the global trading system and their participation thereof, particularly in promoting trade as an engine of growth to further enhance economic and human development in the ESCWA region. This has become one of the key policy objectives to promote their integration into the Multilateral Trading System (MTS) as they shift towards more liberal trade regimes. This is evident in the increasing number of member countries ratifying, signing and joining bilateral, regional and multilateral trade agreements in light of the sluggish progress of the Doha round of negotiations. ESCWA member countries have notified 17 Regional Trade Agreements (RTAs) to the WTO as of April 2008, as shown in Table 9.1. Among the other regional trade and economic arrangements and Free Trade Agreements (FTAs), which have been put into force without notifying the WTO, are the Economic Cooperation Agreement between the European Union (EU) and the Cooperation Council for the Arab States of the Gulf (GCC)[4] in 1988, GCC-Syria FTA (2005) and US-Oman FTA (2005). The agreements that are currently being negotiated are mostly FTAs and are being conducted by the GCC with the following regional organizations: the EU,[5] Azerbaijan, Australia, China, India, Japan, Jordan, Mercosur, New Zealand, Pakistan, Republic of Korea, Singapore and Turkey. Ongoing bilateral trade agreements include the United States-United Arab Emirates, Bahrain-Thailand and Bahrain-Iran FTAs.

Most of ESCWA member countries are members of several regional trade and economic arrangements, including the Greater Arab Free Trade Area (GAFTA) and the GCC Customs Union. GAFTA came into force in 2005. According to the Agreement, all tariffs on commodities of Arab origin were eliminated, leading to an increase in

TABLE 9.1. Concluded FTAs in Western Asia as Notified to WTO (1977–2008)

Year	Name of Agreement	FTAs Notified to WTO by type of Notification (a)		
		GATT Art. XXIV	GATS Art. V	Enabling Clause
1977	EC-Syrian Arab Republic	√		
1997	EC-Palestinian Authority	√		
1998	Greater Arab Free Trade Area	√		
1999	EFTA-Palestinian Authority	√		
2001	United States-Jordan (b)		√	
2002	EFTA-Jordan	√		
	EC-Jordan	√		
2003	EC-Lebanon	√		
	GCC (Customs Union)			√
2004	EC-Egypt	√		
2005	Jordan-Singapore (b)		√	
	Turkey-Palestinian Authority	√		
2006	United States – Bahrain	√		
2007(c)	EFTA-Egypt	√		
	EFTA-Lebanon	√		
	Egypt-Turkey			√
	Turkey-Syrian Arab Republic	√		

Notes: (a) General Agreement on Tariffs and Trade (GATT) Article XXIV covers the formation of customs unions and free trade areas in merchandise trade; Article V of the General Agreement on Trade in Services (GATS) covers agreements in services; and the Enabling Clause deals with trade in goods between developing countries.
(b) EIA: Economic Integration Agreement (EIA): "Agreements that facilitate international trade and cross-border movement of the factors of production. EIAs may address one, some or all these types of economic transactions in various combinations, with trade being the central component, and foreign investment activity one possible ingredient."
(c) Data as of 15 April 2008.
Source: WTO (2008)

trade among the Arab countries. During 2002 and 2006, intra regional Arab exports and imports witnessed steady and high annual growth rates. The share of intra-Arab trade to total world trade increased to 11.3 per cent in 2006 from 9.3 per cent in 2001[6]. Likewise, intraregional trade among members of the GCC Customs Union[7] experienced an increase in both its imports and its exports, registering an increase of 304.8 per cent in its exports during the period 1994–2005[8]. As a regional bloc, the GCC launched their joint common market in January 2008 in order to "... create one market... raising production efficiency and optimum usage of available resources and improving the GCC's negotiating position in international economic forums.[9]" GCC countries also plan to introduce a common currency by 2010.

Other arrangements include the Agadir Agreement, which was signed in 2004 and entered into force in 2006 between two ESCWA member countries (Jordan and Egypt) and two non-ESCWA Arab member countries (Tunisia and Morocco). The agreement seeks to establish a free trade area among its members with the final objective of setting up of a joint European- Mediterranean free trade area.

Trade liberalization has played an important role as an engine for growth in the region. As a result, member countries are experiencing changes in their trade regimes. They are all moving towards a more liberal trade regime, trying to utilize trade as an engine for growth. Some have adopted one form of export promotion strategy to replace the import substitution strategies largely due to adhering to commitments of the Uruguay Round, resulting from their accession to the WTO or ratifying RTAs.[10]

Another important indicator of trade liberalization in the regional and multilateral context is the liberalization of trade in services. The degree of the liberalization of trade in services, as reflected in the General Agreement on Trade in Services (GATS) commitments of ESCWA member countries, differs considerably from one country to another. The number of commitments in services undertaken by countries that are relatively new WTO members is much higher than those countries which became WTO members at an earlier date. The number of services trade commitments made under GATS for those ESCWA member countries that joined the WTO after 2000, Saudi Arabia made 120 commitments, Jordan made 110 commitments and Oman 98. Those that joined earlier in 1995 and 1996, Bahrain had made 22 commitments in trade in services, Egypt 44, Kuwait 61, Qatar 46 and the United Arab Emirates 46[11].

Trade facilitation is also playing an important role in the region, whereby trade facilitation mechanisms adopted under RTAs usually take multifaceted measures ranging from mutual recognition and certification to standardization of documentation requirements, simplification of trade procedures, border coordination and cooperation, enhanced connectivity and use of ICT to policy coordination, among others. The value of basing these measures on international standards will facilitate moving from implementation within a region to multilateral implementation. The use of international standards will also facilitate implementing trade facilitation across country groupings and between countries that are signatories to different RTAs. Furthermore, the region is increasingly recognizing that trade facilitation is key to realizing the benefits of trade liberalization through the simplification and harmonisation of trade procedures by reducing transport costs, improving port facilities, establishing efficient customs regimes, having transparent and coordinated regulations, improving information quality and information technology infrastructure and facilitating market access. However, it must be noted that the degree of liberalization differs among ESCWA member countries – GCC countries in general seem to be more liberal when compared to non-GCC members.

9.3. MAPPING AID FLOWS TO THE ESCWA REGION

According to the Official Development Assistance (ODA) flows of the Development Assistance Committee (DAC) at the Organisation for Economic Co-operation and Development (OECD), the DAC list categorizes countries according to their level of income. According to their list, Lebanon and Oman are considered upper middle-income countries (UMICs); Iraq, Egypt, Jordan, Palestine and the Syrian Arab Republic are categorized as lower middle-income countries (LMICs); and Yemen is

TABLE 9.2. ESCWA Recipients of ODA and Selected Indicators, 2006 (In USD Million)

ESCWA Members (selected)	2006
Lebanon	707
Palestinian Adm. Areas	1,449
Iraq	8,661
Jordan	580
Egypt	873
Syria	27
Bahrain	-
Saudi Arabia	25
Oman	35
Yemen	284
Middle East, Regional	2,056
Total ODA to ESCWA Region	**14,697**
Total AFT (TRTA)	360
Multilateral TRTA	116
Bilateral TRTA	244
Total OECD ODA	103,900
ESCWA Share of Total ODA (%)	**14**

Source: ESCWA calculations, based on WTO/OECD Database (2008)

the only least developed country (LDC) in the ESCWA region and the only ESCWA country already included in the Integrated Framework (IF) programme, which is a comprehensive AfT approach for LDCs.[12] In terms of gross domestic product (GDP) per capital and growth rate, Yemen has the lowest in the region. The GDP per capita was US$ 607 and the growth rate was 3 per cent in 2006. The remaining gulf countries have graduated from the DAC recipient list. Bahrain graduated the list in January 2005 and Saudi Arabia left in 2007. All of these countries have received some form of development assistance from developing countries.

Based on OECD data, nine out of 13 ESCWA member countries received 14 per cent of the world's total ODA in 2006, amounting to slightly over USD 14 billion (Table 9.2)[13]. The three ESCWA member countries that did not receive aid in 2006 are Bahrain, Kuwait, Qatar and the United Arab Emirates. Acknowledging the inherent difficulties in monitoring AfT, due to the wide scope of aid that can be classified as AfT, the OECD estimates that in 2005, ODA commitments USD 106.8 billion to USD 103.9 billion in 2006, down by 5.1 per cent from 2005. ODA was exceptionally high in 2005 due to large debt relief operations, especially in Iraq and Nigeria. Excluding debt relief, ODA fell by 1.8 per cent[14]). Using a broad scope of AfT, which includes aid for trade-related infrastructure, building productive capacity, trade-related infrastructure and trade policy regulations, Egypt was the sixth highest recipient of AfT globally, receiving an average of over USD 500 million from 2002–2005. No other ESCWA member Country ranked in the top 25 recipients of AfT

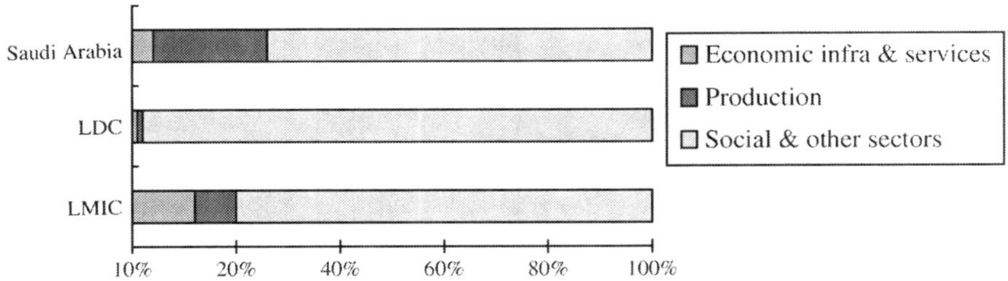

FIGURE 9.1. Sectoral Distribution of net ODA (2004–2006) (*Source*: OECD (2007))

(OECD/WTO, 2007). Using a limited scope of AfT, the ESCWA region as a whole received only USD 36 million as trade-related technical assistance, or TRTA[15] , of which USD 24 million was provided as bilateral aid and USD 12 million as multilateral aid in 2006. TRTA represented 2.5 per cent of the total ODA received in the ESCWA region in 2006, is of marginal significance and does not commensurate with the necessary funds to address the prevailing needs of the region.

In the past, ODA concentrated on social infrastructure rather than economic infrastructure and services and production between 2004 and 2006. The majority of aid went to the social sector,which includes water, sewage, education, health, housing and rural development, as seen in Figure 9.1. In terms of how much the ESCWA sub regions received for trade-related technical assistance, the UMICs received the most, followed by the LMICs and the sole LDC in the region. Graduating the DAC list in 2007, ODA to Saudi Arabia concentrated on social infrastructure as well. However, it also utilized aid in production between 2004 and 2006.

9.4. THE ESCWA REGION AS A DONOR OF AID

The ESCWA region has not only been a recipient of aid. Member countries and regional development banks have historically played an important role in providing aid to developing countries both within and outside the region.

A total of USD 129 billion was given as aid by the Arab countries from 1970 to 2006. In 2006, the ESCWA region provided USD 3.3 billion in aid. Most of this aid came from the oil-rich countries[16]. In 2006, the largest contributor was Saudi Arabia, which provided 66 per cent of the aid, followed by Kuwait at 18 per cent and Qatar at 15 per cent.

Arab aid is characterized by the donor's generosity. Back in the 1970s, Arab aid averaged 4.7 per cent of their combined Gross National Income (GNI). Though the bulk of the aid goes to Arab countries, it is volatile and unpredictable due to the fluctuation of the price of the oil. The sources of aid are either formal or informal bilateral aid. The Coordination Group of Arab Donors coordinates policies and practices among its members for maximum operational efficiency. Members comprising

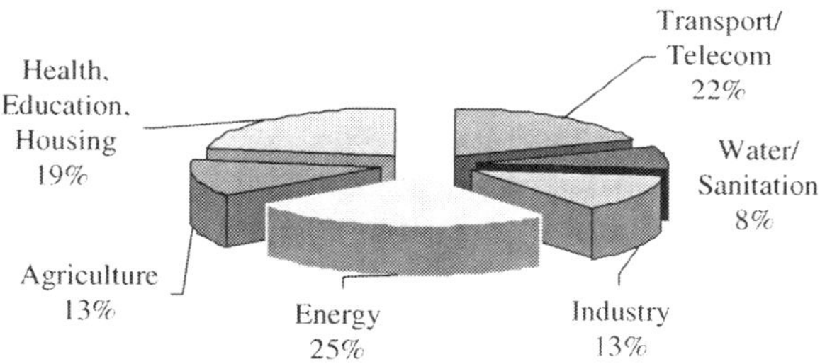

FIGURE 9.2. Sectoral Distribution of Aid from the Arab Region (2006) (*Source*: Unified Arab Economic Report, Arab Monetary Fund (2007))

of funds or agencies are Abu Dhabi, Kuwait, Saudi Arabia, OPEC and Arab Funds, Arab Bank for Economic Development in Africa (BADEA), Islamic Development Bank and Arab Gulf Programme for United Nations Development Organizations (AGFUND). Another characteristic of Arab aid is that Arab donors have always adhered to the principle of non-interference in recipient country policies. Arab aid is unconditional. It is given to help recipients "set up policies for their economic and social development of their own free will outside of political and economic pressures." The recipient countries are supposed to determine their own economic and development policies, and therefore receive aid without any conditions attached.

According to the 2007 Unified Arab Economic Report of the Arab Monetary Fund (AMF), more than half of Arab aid was intra regional aid. Egypt was the largest recipient receiving 26 per cent of the total amount, followed by Iraq (22 per cent), Palestine (7 per cent), Jordan (6 per cent), and Yemen (4 per cent) during the cumulative period from 1990 to 2006. In other regions, Asia received 22 per cent of the Arab aid, Africa received 15 per cent and Latin America 1 per cent in 2006.

The sectoral distribution of Arab aid in 2006 primarily supported the energy sector, the transport and telecommunications sector, and to a lesser extent the agriculture and industrial and water and sanitation sectors, as seen in Figure 9.2. Apart from the energy sector, Arab aid is mostly channeled into the transport and telecommunications, agriculture and industrial sectors, which represented 48 per cent of cumulative ODA by the end of 2006 in the Arab region.

9.5. ADDRESSING SUPPLY-SIDE CONSTRAINTS AND MARKET ACCESS CHALLENGES IN THE ESCWA REGION

In order to better integrate into the world economy and to make trade an effective engine of growth and development, ESCWA member countries need to further promote their capacity to produce competitively in order to penetrate foreign markets. To this end, AfT can be utilized to further address and overcome a variety of regional-specific constraints as well as to continue to improve trade negotiation and policy formulation skills that will allow ESCWA member countries to further benefit from the multilateral trading system. AfT can also be used to improve existing infrastructure by, for example improving road networks to increase the speed that products reach ports and borders and enhance trade through improved use of information and communication technology (ICT) to meet product standards, technical barriers and certifications. AfT could support research and development (R&D) initiatives, technology transfer and assist countries to meet international regulatory standards, such as food safety and animal and plant health measures (e.g [SPS][17] Sanitary and Phytosanitary Measures). At the same time, AfT can be used in order to improve the competitiveness of exports to simplify and harmonize trade procedures even more, such as customs clearance procedures, non-tariff barriers and impediments to labour mobility.

Among some of the existing region-specific supply-side constraints which could benefit from further improvement are customs regimes and standards, non-tariff standards, mainstreaming trade policies in national strategies and enhancing the competitiveness of exports of goods and services.

9.5.1 MORE EFFICIENT INSTITUTIONS

Just as the quality of laws and regulations is important, the efficiency and governance of the institutions that enforce them are equally important[18]. To implement trade liberalization policies effectively, greater attention must be given to assist the development of competent institutions in member countries to further improve the smooth flow of export-import procedures, such as enhancing the efficient skills of human resources at customs and port authorities and more accountable policies related to customs clearance. Also, the quality assurance infrastructure in the region, such as testing and metrology labs and affiliated accreditation agencies, could be further improved as they play an important role in enabling companies to actively participate in international trade. Thus, institutional capacity can affect trade costs if customs procedures, inspections and certifying bodies operating and functioning as efficiently as they are not should.

An example of this major impediment to produce goods and services competitively and the ability to get them to international markets at a reasonable cost can be understood by comparing the procedural requirements for trade vis-à-vis other regions, as seen in Table 9.3 On average, it takes 29 days to process exports from the ESCWA region as compared to 10 days among OECD member countries. Seven documents are needed to process exports in the ESCWA region, only five to process import in the ESCWA region compared to documents for the OECD countries. Similarly, it takes 32 days as compared to 10 days in OECD countries. Eight documents are needed for ESCWA region as opposed to 5 documents for the OECD countries. Processing exports and imports in the ESCWA region takes about 20 days longer than in OECD member countries, and requires more paperwork. The ESCWA region lags

TABLE 9.3. Procedural Requirements for Exporting/Importing, Selected Regions

Region	Processing time (Days)		Document (Number)	
	Export	Import	Export	Import
ESCWA	28.5	32	7.1	8.1
Latin America & Caribbean	22.2	25.8	7	7.6
East Asia & Pacific	24.5	25.8	7.0	7.5
OECD	9.8	10.4	4.5	5

Source: Adapted calculations from the World Bank (2007)

far behind in the procedural requirements for exporting and importing compared to the developed countries.

9.5.2 Infrastructure

Strong infrastructure is an important means to overcome supply-side constraints in the ESCWA region. A more efficient transport infrastructure can prevent obtaining access to foreign markets and international ports. Better storage facilities can increase inventory costs and insufficient energy and water supplies can disrupt production or increase costs. However, superior road and port facilities alone are not the only impediment to increase trade. Infrastructural improvements have to go hand in hand with good governance. The value of infrastructure projects is easily eroded by inadequate economic policies or inefficient customs services.

9.5.3 Trade Facilitation

Though trade facilitation has played an important role in the ESCWA region and has become a main concern for national trade liberalization policies. Though trade facilitation has been promoted in the region, overall trade facilitation measures still need to be improved in order to simplify and harmonize trade procedures, such as customs clearance procedures, non-tariff barriers and impediments to labour mobility. Trade facility in the region could benefit from improvements in:

- Trade-related infrastructure, including storage facilities;
- Knowledge of foreign market opportunities;
- Customs administration;
- The capacity of the private sector to engage in more effective marketing and promotion[19].

9.5.4 Private Sector Involvement

The private sector in the ESCWA region has been quite active in the past. However, more needs to be done to enhance their participation in national policymaking processes and the multilateral trading system in order to enable them to reap the full benefits of the trading system. One possible mechanism is to establish an institutional structure to link the private sector with national and regional policymakers to help them identify their needs and priorities, coordinate a common position, formulate a strategy to advance their trade-related agenda and strengthen their voice in the multilateral trading system. To this end, national and regional policy-oriented think tanks could be created to tackle trade-related issues of importance to the private sector and help to build bridges between business community and policymakers.

To address these constraints, the participation of various stakeholders such as national governments, civil society and the private sector, is critical in efforts to mainstream AfT into national policies and leverage trade as an engine for growth. The potential of AfT to address some of the regional constraints to trade is significant; however, recipient countries must also work to create a conducive environment for aid to be used effectively and efficiently. The following measures could be considered to achieve this:

i. A comprehensive national economic and trade strategy could be developed to improve competitiveness and market access. This national trade strategy should outline specific objectives and targets suited to the country in question as individual countries have their own needs and requirements.
ii. Trade policies should be mainstreamed into national development agendas thereby constituting a roadmap towards the achievement of trade-related objectives, including regulatory reform, institutional reform and re-alignment, capacity-building, developing public awareness and investing in new or existing infrastructure.
iii. Trade-related needs and priorities should be identified when formulating a short and long-term workplan towards these trade-related objectives and address specifics on how international aid can be utilized.
iv. The resources needed for the different required inputs should be identified. Some countries may already be in a position to undertake certain activities on their own, and this practise will bring to light those areas where external support will be required to maintain an effective pace. In other words, this is the stage where trade-related assistance or AfT can first be discussed in a comprehensive manner
v. Domestic coordination mechanism among the various stakeholders, including the private sector could be improved. Once a national plan is in place, a national institutional body is necessary to oversee AfT activities. This institutional body should have direct access to an executive body to ensure the smooth and efficient implementation of the overall decision-making process.

All of these steps would require Governments to efficiently build their national statistical capacity to collect and analyze AfT-related data. Thus, the opportunities that a successful AfT initiative would bring to the ESCWA region include, but are not limited, to:

- Ensuring that the trade-related needs of ESCWA member countries are met through AfT, which is aligned to national development strategies.
- Increasing the amount of aid entering the region to assist them in meeting their priority needs.
- Providing and harmonizing procedures a much more effective coordination scheme among donors procedures for donors.
- Enhancing the transparency and efficiency of AfT through a systematic reviewing, monitoring and evaluation process.
- By leveraging trade as an engine for growth through the AfT initiative, it would promote further integration and participation of the ESCWA region into the multilateral trading system.

9.6. ENSURING A SUCCESSFUL OUTCOME OF THE AFT INITIATIVE IN THE ESCWA REGION: POLICY IMPLICATIONS

The global debate on AfT should include all regions covered by the United Nations regional commissions, including the ESCWA region, as data show that the ESCWA region is not only a recipient of aid from developed countries but it is also a donor of aid to other regions. The ESCWA region should play a greater role in the AfT initiative, both a potential recipient and donor. At present, AfT is not a priority area for most of the countries in the region, particularly crisis-stricken countries, such as Iraq and Palestine. Stronger efforts are needed to operationalize this initiative at the national and regional level in order to enhance economic development in individual countries and as the region as a whole: all stakeholders have an important role to play in ensuring a successful outcome of the AfT initiative in the ESCWA region, and promoting greater integration into the multilateral trading system.

The ESCWA region is unique in that it consists of 14 member countries and many of its countries have large oil reserves. However, is has a modest institutional and infrastructural base and existing national and regional mechanisms to formulate policies and oversee and coordinate activities are at times inefficient. While the public sector is dominant in the economies of the region despite efforts for liberalization and privatization, and an emerging private sector, the ESCWA region has a lot of potential in nominal terms. The region's gross domestic product (GDP) share of the world total increased by 0.34 per cent between 2001 and 2006. The balance of trade increased by 0.76 per cent in 2006 from 2001 of the total world trade. Foreign Direct Investment (FDI) in the region increased significantly by 3.43 per cent of the total world FDI in the same period. The most noteworthy increase is the real GDP growth rate in terms of the world total, where it increased about 4 per cent. In this regard where AfT is concerned, its definition should be clarified and streamlined, avoiding the "one-size-fits-all" approach. As the development of AfT should be consistent with the Paris Declaration on Aid Effectiveness,[20] it should also be in line with the broader trade objectives and the principle of South-South cooperation. Hence, the current mechanisms for the delivery of AfT should be more demand-driven and aligned to national development strategies.

The majority of the countries in the region are in much need of trade-related technical assistance to upgrade their domestic trade environment. However, assistance is needed to ensure that AfT promotes development. Assistance in identifying their priorities; monitor and evaluate quantitative indicators of donor and recipient performance; recipient country ownership; alignment between donor support and partner country development strategies and capacity; harmonization of donor activities; results; and mutual accountability are some of the areas as mentioned in the Paris Declaration.

Up until now, there have been numerous programmes tailor-made to specific needs. But these have been prepared mostly at the spur of the moment to deal with specific gaps in economic infrastructure and services. A short-term comprehensive trade policy framework has unfortunately been the norm in most of the ESCWA member countries. The region should now seriously consider re-evaluating its needs, both as individual countries and collectively as a region, *vis-à-vis* AfT. Furthermore, a long-term, inclusive and effective AfT framework is needed in the region with the assistance of various stakeholders.

9.7. ESCWA'S ROLE DEFINED: THE PAST, PRESENT AND FUTURE

As one of the five United Nations regional commissions, ESCWA has played an important role in assisting member countries and Arab countries in trade-related technical assistance and capacity building in the past. ESCWA has assisted its member countries to build their capacity in: (i) transboundary flows of goods and services; (ii) negotiation and accession skills for the WTO and other trade agreements; (iii) implementation of the Monterrey Consensus that emanated from the Financing for Development Conference; and (iv) managing the opportunities and challenges posed by globalization.

ESCWA aims to facilitate trans boundary flows of goods, services, persons and capital in order to meet the challenges posed by the trend towards globalization. It focuses on integrating regional markets to cope with emerging economic blocs, with emphasis on facilitating trans boundary flows of goods, services and persons; responding to the trends of global trade liberalization in order to minimize potential threats and seize the opportunities offered by analyzing the evolving trends of the new regional and global trading systems, with special emphasis on the Arab Free Trade Area (AFTA) and the WTO. It also assists member countries in improving export performance, taking advantage of the international competition, and enhancing regional development through encouraging domestic as well as foreign investments. These are achieved by conducting studies and advising the member states on the latest world trade issues and the economic and market considerations under the WTO, promoting policies and programmes aimed at strengthening the process of economic cooperation and integration between the member states, conducting studies on member states policies on intra regional and international movement of people, capital and trade and facilitating regional cross-border trade and investment, conducting comparative studies of the impact of fiscal and monetary polices on the development of regional cooperation, conducting labour demand and supply analysis and studying intra regional labour flows, studying the existing and projected needs of the tourist industry and of transportation infrastructures in the region; and providing member states with technical assistance on current world trade issues.

Hence, ESCWA fulfils its mandate by undertaking a combination of normative, analytical and operational activities in the field of economic development and integration. Technical assistance focuses on improving the capacity of countries through policy advice and capacity-building activities through advisory services, technical cooperation projects and workshops and training courses, to individual countries or at a regional/sub-regional level.

Emphasis is also placed on strengthening the negotiation skills of member countries in respect of the WTO negotiation process, continuing to increase their understanding of the opportunities and challenges presented by WTO as well as regional integration agreements, defining obstacles and barriers to participation in the multilateral trading system and organizing awareness drives on the role of the Arab free-trade areas and other bilateral agreements. Support is provided in enhancing regional development by analysing existing credit and investment policies and formulating specific recommendations for the improvement of existing investment laws, regulations and incentives in order to increase significantly the attractiveness of member countries to domestic and foreign investors, with particular emphasis on interregional capital flows and the return of capital invested outside the region.

Among other tools adopted by ESCWA to fulfill its mandate and objectives are:

a. Ministerial coordination meetings, such as the Arab Ministerial Meeting in Preparation for the WTO Ministerial Conference;
b. Intergovernmental and expert group meetings to follow up the results of the WTO 6th Ministerial Conference;
c. Seminars and training/capacity-building workshops on relevant topics such as agricultural trade, WTO negotiations, trade in non-agricultural goods, trade in services, intellectual property rights, trade facilitation, anti-dumping, movement of natural persons and single window system;
d. Analytical and normative work, such as the Annual Review of Developments in Globalization and Regional Integration in the Countries of the ESCWA Region

and Movement of Natural Persons in the ESCWA Region, Aid for Trade and South-South Cooperation;

e. Implementing and monitoring international agreements in the ESCWA region, such as adoption of the Agreement on International Railways in the Arab Mashreq and Integrated Transport System in the Arab Mashreq (ITSAM), adoption of the Memorandum of Understanding on maritime transport cooperation in the Arab Mashreq;
f. Establishment of the Technical Committee on Liberalization of Foreign Trade and Economic Globalization in the countries of the ESCWA region;
g. Responding to member country requests for technical assistance and advisory services on multilateral trade issues and negotiations;
h. Maintenance and regular updating of ESCWA's WTO related website in Arabic on regional integration and responding to globalization on the Internet;
i. Networking and knowledge-sharing, as well as field projects, such as the development account, "Interregional partnership for promoting trade as an engine for growth through knowledge management and ICT," where the five regional commissions are actively involved in;
j. Cooperation with other United Nations agencies and entities, such as the Joint Project on the Accession of the Syrian Arab Republic to the WTO, in collaboration with the United Nations Development Programme and UNCTAD and other regional organizations, such as League of Arab States and Arab Labour Organization.

ESCWA therefore is well positioned to continue assisting in the implementation of the AfT initiative, whereby regional initiatives on the ESCWA level would go hand in hand with national initiatives by:

1. Promoting and supporting future debate on AfT by participating in the strategies set forward, such as National and subregional reviews, regional AfT networks and regional and global monitoring;
2. Assisting member countries, donors and recipients in surveying and identifying their AfT priority needs based on their trade policies;
3. Assessing the impact and efficiency of customized AfT funding has on productive capacity and trade performance of recipient member countries;
4. Continuing to provide target-oriented technical assistance for member countries, built on best practices of other United Nations regional commissions and in cooperation with other stakeholders and public-private partnerships (PPPs), on both the regional and international levels;
5. Encouraging dialogue between member countries, donors and relevant regional organizations on issues related to AfT, trade facilitation, among others.

ESCWA is integrating AfT in its overall development strategy and programming and will continue to systematically do so in its future work. This will be done particularly within the framework of identifying cross-cutting issues, such as capacity-building for trade focusing on mainstreaming trade in national development strategies; promoting PPPs; strengthening the private sector, especially small and medium-sized enterprises (SMEs); and promoting South-South cooperation in trade, FDI, infrastructure development, ICT, water management, energy, tourism and agricultural and rural development.

Furthermore, ESCWA could become the focal point for the ESCWA region with regards to AfT-related issues. ESCWA should become the centre for AfT-related issues in the ESCWA region, in close collaboration with other international organizations and PPTs. In coordination with all relevant agencies and organizations on a common theme and budgetary allocation, ESCWA would be able to bring together regional donors and recipients to address the issue of AfT in the region and hold national or subregional reviews and then a regional review on AfT to gather and coordinate financial resources and gather statistical data on AfT in the region and develop a database for AfT work in the ESCWA region. ESCWA would work with international organizations, such as the World Bank, private and public sectors and non-governmental organizations, to forge partnerships in areas of technical assistance and capacity-building on AfT issues. Actively participate and encourage ESCWA member countries to participate in the global debate on AfT, collaborating with the various organizations, institutions, agencies and others. ESCWA could also look into encouraging the establishment of Regional Aid for Trade networks – comprised of key stakeholders (lead donors and regional funds, etc.) – to assist member countries in identifying priorities, mainstreaming trade policies into their national agendas and mobilizing resources.

Bearing in mind the importance of the role of AfT in promoting trade as an engine for growth to further economic development, capacity of ESCWA member countries to better negotiate and implement subregional, regional and multilateral trade agreements for increasing intraregional and international trade should be enhanced, while raising awareness of the ESCWA member countries of the importance of prioritizing AfT in their national strategies and agendas. To further promote trade facilitation, the capacity of member countries to implement the various components of ITSAM to facilitate the flows of goods and passengers among countries of the region should be enhanced. Furthermore, member countries should be encouraged to adhere to the general guidelines, scope and definition of the AfT Initiative, which emerged as a significant part of the WTO Hong Kong Ministerial Declaration in 2005. There is a necessity to build awareness and interest of the AfT Initiative in the region, while mainstreaming AfT and trade-related policies into national development strategies and push forward with implementing and operationalizing AfT. Hence, ESCWA will continue to assist member countries in taking the appropriate steps and measures in promoting trade as an engine for growth in the region by continuing to integrate AfT in its overall development strategy.

NOTES

[1] The Sixth WTO Ministerial Conference was held in Hong Kong in 2005, where WTO Members mandated work on Aid for Trade in paragraph 57 of the Hong Kong Ministerial Declaration (http://www.wto.org/english/thewto_e/minist_e/min05_e/final_text_e.htm#aid_for_trade), whereby a task force was established to provide recommendations on how to operationalize Aid for Trade.

[2] For further information on WTO's work on AfT, including reports and reviews, see http://www.wto.org/english/tratop_e/devel_e/a4t_e/aid4trade_e.htm.

[3] ESCWA is made up of 14 member countries: Bahrain, Egypt, Iraq, Jordan, Kuwait, Lebanon, Oman, Palestine, Qatar, Saudi Arabia, Syrian Arab Republic, United Arab Emirates, Yemen and the Sudan.

[4] The Cooperation Council for the Arab States of the Gulf,otherwise known as the Gulf Cooperation Council (GCC), was established in 1981 and is an economic and political policy-coordinating forum for six member States (Bahrain, Kuwait, Oman, Qatar, Saudi Arabia and the United Arab Emirates (UAE)). The GCC cannot impose trade policies upon its member states. Each is free to pass and enforce its own trade laws. However, there has been growing cooperation among GCC member states on issues such as customs duties, intellectual property protection, standards-setting and intra-GCC investments. For further information, see http://www.gcc-sg.org.

[5] See EU-GCC relations at (http://ec.europa.eu/external_relations/gulf_cooperation/intro/index.htm).

[6] ESCWA 2007. Annual Review of Developments in Globalization and Regional Integration in the Arab Countries, (in Arabic), p. 32, table 14. United Nations New York, (E/ESCWA/GRID/2007/14).

[7] The GCC Agreement establishing the Customs Union contains provisions related to trade in goods that are relevant to the Members of the WTO. The Agreement establishes a customs union as portrayed in Article XXIV of the GATT. By establishing a customs union, they adopted unified customs laws and procedures, single point-of-entry with internally free movement of goods and treatment of goods as national origin within the GCC. Duties and other restrictive regulations of commerce have been eliminated on substantially all trade between the signatories, and a uniform external tariff has been adopted by the parties to the agreement. Source: World Trade Organization, WT/REG222/N/1, 20 November 2006 (06-5577).

[8] For further information, see the GCC's homepage (http://www.gcc-sg.org) and the Federation of GCC Chambers homepage (http://www.fgccc.org).

[9] Final Communiqué of the 28th GCC Summit (http://www.gcc-sg.org/index.php?action=Sec-Show&ID=257) in Arabic.

[10] Out of the 14 ESCWA member countries, eight are WTO members (Bahrain, Egypt, Jordan, Kuwait, Oman, Qatar, Saudi Arabia and United Arab Emirates). Iraq, Lebanon and Yemen currently have observer status and their accession to the WTO is under consideration. Palestine and the Syrian Arab Republic are not yet members. http://www.wto.org/english/thewto_e/whatis_e/tif_e/org6_e.htm

[11] Source: Country Profiles, WTO Database (April 2007).

[12] "The Integrated Framework for Trade-Related Technical Assistance to least-developed countries (IF) is a multi-agency, multi-donor programme that assists the LDCs to expand their participation in the global economy and enhance their economic growth and poverty reduction strategies. The IF programme was first mandated by the WTO Singapore Ministerial Conference in December 1996. The participating agencies are IMF, ITC, UNCTAD, UNDP, World Bank and the WTO." See also http://www.integratedframework.org/

[13] At the time of the study, the Sudan was not yet a member of ESCWA. Therefore the calculations in Table 9.2. do not include the Sudan.

[14] See: http://www.oecd.org/document/17/0,3343,en_2649_34447_38341265_1_1_1_1,00.html

[15] Trade-Related Technical Assistance/Capacity-Building (TRTA/CB) is defined by the OECD as activities aimed at enhancing the ability of recipient countries to: (a) formulate and implement a trade development strategy and create an enabling environment for increasing the volume and value-added of exports, diversifying export products and markets and increasing foreign investment to generate jobs and trade, (b) stimulate trade by domestic firms and encourage investment in trade-oriented industries and (c) participate in and benefit from the institutions, negotiations.

[16] The effectiveness of such an amount has yet to be analyzed further in detail in terms of its impact on national, regional and international trade, trade facilitation measures, trade policy implementation and so on.

[17] For further information on sanitary and phytosanitary measures, see http://www.wto.org/english/tratop_e/sps_e/sps_e.htm.

[18] "Addressing Supply-side Constraints and Capacity Building," Note by the Secretariat, E/ESCAP/SCITI/3; 2 September 2004.

[19] "Aid-for-Trade: A Perspective from the ESCWA Region," 2008 ESCWA (unpublished).

[20] The Paris Declaration on Aid Effectiveness was endorsed in 2005 and is an international agreement to OECD member states vowed "to continue to increase efforts in harmonization, alignment and managing aid for results with a set of monitorable actions and indicators."

REFERENCES

Arab Monetary Fund (2006 and 2007). Unified Arab Economic Report (Arabic).

ESCAP (2007). Trade and Investment Issues: Aid For Trade And Public-Private Partnerships, Note by the secretariat, Committee on Managing Globalisation, Fourth Session.

ESCWA (2007). Economic Trends and Impacts: Foreign Aid and Development in the Arab Region, Distr. General, E/ESCWA/EAD/2007/14 January 2007 (English), Issue No. 4.

ESCWA (2008a). Promoting Aid-for-Trade in the ESCWA Region, Presentation made by ESCWA at UNCTAD XII, Accra, Ghana, 23 April 2008.

ESCWA (2008b). Report of the UN Secretary-General on Regional Cooperation in the economic, social and related fields at ECOSOC.

ESCWA (2008). Aid-for-Trade: A Perspective from the ESCWA Region (unpublished).

OECD (2006). The Development Dimension, Aid for Trade: Making it Effective.

OECD (2007a). After Gleneagles: What Role for Loans in ODA?, Policy Brief No. 31.

OECD (2007b). Aid Effectiveness: 2006 Survey on Monitoring the Paris Declaration, Overview of the Results. *OECD Journal on Development*, 8(2).

OECD (2007c). Final ODA Flows in 2006, DCD/DAC/RD(2007)15/RD2.

OECD/WTO (2006). 2006 Joint WTO/OECD Report on Trade-Related Technical Assistance and Capacity Building.

OECD/WTO (2007). Aid for Trade at a Glance 2007, 1st Global Review, p. 29.

World Bank/IMF (2007). Aid for Trade: Harnessing Globalisation for Economic Development.

UNECA

CHAPTER 10

BUILDING AFRICA'S SUPPLY CAPACITIES AND COMPETITIVENESS THROUGH AID FOR TRADE

10.1. INTRODUCTION

This chapter discusses the Aid for Trade (AfT) initiative and its location in dealing with Africa's trade challenges[1]. The AfT is evolving towards being a core area within the broad development finance architecture. Essentially, the AfT idea as will be seen in this chapter borders on ring-fenced resources targeted to helping developing countries integration into the global trading system. This paper starts with a discussion of the trade challenges that Africa faces, both internally and externally. These challenges require a response such as that envisaged in the AfT as can be seen from its rationale and its scope which are discussed in this chapter.

10.2. TRADE AS A DEVELOPMENT PILLAR

It is well accepted that trade is an important development pillar in Africa. The current engagement of African countries in multilateral and bilateral trade negotiations, and the regional integration initiatives in the continent, is evidence of the central role attributed to trade in catalysing growth and hence development. A lot of countries have also shown interest in mainstreaming trade in national development strategies as recognition to the role trade could play in their development.

Beyond the policy arena, there has been a lot of debate, driven by the experience of African countries, regarding how trade policy itself could be used to enable trade play this crucial developmental role. There are several studies that capture concisely the elements of this debate, which revolves around the question of the beneficial effects on growth and poverty reduction of trade liberalisation. What would be useful at this juncture though is to point out that there appears to be a consensus that developing countries like those in Africa neither need neo-liberal trade policies as in the 80s nor interventionist ones like in the 70s. Rather they need proactive and pragmatic trade policies that recognise individual countries development visions and circumstances. Notwithstanding this debate, the most important point is that trade and trade policy is at the mainstream of today's economic policy debate.

Yet, the importance of trade and development is a more urgent issue in Africa than in any other region. This is because the region still experiences a very low level of integration in the global trading system. To put this picture in perspectives, today, Africa's share of global exports especially in merchandises remains historically low. The continent's share in global exports in 2005 was a mere 2.8 per cent,

P. De Lombaerde, L. Puri (eds.), *Aid for Trade: Global and Regional Perspectives*,
United Nations University Series on Regionalism 2,
DOI 10.1007/978-1-4020-9455-2_11, © Springer Science+Business Media B.V. 2009

roughly equivalent to its 1991 value and less than half its peak value of 6 per cent last achieved in 1980. The region has also been unable to achieve the desired outcomes in regional integration at the continental level. These two developments have intensified the focus on trade in Africa.

And as noted in ECA's Economic Report on Africa 2007, over the past 6 years, world merchandise exports experienced an average growth rate of 10.4 per cent. Over this period, Africa performed better, increasing its exports by 16 per cent on average annually. However, a closer look at this export performance reveals that the rapid increase in export is concentrated in oil-exporting Sub-Saharan African countries. Thus, it is the price movements, rather than major structural changes in Africa's production and export patterns that have been driving the recent rapid exports growth. Weak supply-side capacities and lack of diversification continue to haunt Africa's endeavours to increase its share in the global trade. As a result, like in the past, Africa is still vulnerable to shifts in international commodity prices, and particularly in changes in prices of oil.

10.3. AFRICA'S TRADE CHALLENGES

Given the importance of trade as an important pillar in Africa's development, the question that comes to mind is why the region has not been able to harness this potential? The main answer given to this question in this paper is that Africa still faces major trade challenges. By highlighting these challenges, this paper allows for an objective examination of how best to operationalise the AfT initiative in a way that will help it effectively address the African challenges. At this point, it helps to observe that the AfT initiative as espoused in its rationale and defined scope goes beyond the traditional trade capacity building that programmes like the Integrated Framework (IF) and the Joint Integrated Technical Assistance Programme (JITAP) have been focusing on.

Yet, the African trade challenges go beyond these capacity constraints. In the remainder of this section, the two main categories of trade challenges, not related to trade negotiations capacities are outlined. There are both external and internal challenges. The external challenges happen to be exogenous at the same time in the sense that they are given and Africa as a region cannot do anything about them, as they are driven by policies outside its realm. In the case of the internal challenges, there two types, the software related and those to do with the hardware elements of the trading environment. This section will therefore briefly look at each of these challenges.

10.4. THE EXTERNAL TRADE CHALLENGES

These challenges relate mainly to the international trade environment that Africa faces. As a result, they have more to do with the market access difficulties that Africa's exports have to deal with. The main elements of the external trade challenges include:

Tariff peaks: African exports are still faced with a protection pattern characterised by a large number of tariff peaks. These peaks concern mainly products of agriculture, food, textiles, and apparel, which require low- and mid-technologies.

Tariff escalation: Besides the tariff peaks, Africa faces yet another form of market access constraint that is related to tariff escalation. Any value-addition which could lead not only to diversification but to a higher percentage of income benefiting African economies is discouraged by escalating tariffs on semi-processed goods.

Export subsidies: Export subsidies provided by some of the developed countries also affect market access conditions that ultimately face African goods. The outcome of the export subsidies has been shown to be depressed international prices, which result in an unfair playing field, as African exports get "out-competed" in these markets where they could otherwise have done better.

Domestic support policies in developed countries: Domestic support measures leads to overproduction, creating international markets gluts, which then result in international prices collapse that eventually negatively affect African exports. The cotton market is a good example where domestic support measures in some developed countries has led to depressed international prices which make African cotton producers look less competitive.

Non-Tariff Barriers (NTBs): NTBs also pose market access challenges for Africa. The imposition of sanitary and phyto-sanitary measures (SPS) and the Technical Barriers to Trade (TBT) constitute a constraint for Africa in the export of its products to developed countries. The market access constraints related to SPS are more pronounced when one considers the limited capacity for the African countries to meet the measures.

Restrictive rules of origin: One might argue that all the above elements related to external market access challenges for Africa are exaggerated given that the region enjoys a wide array of preferential market access schemes. The Everything But Arms and the African Growth and Opportunity Act quickly come to mind, as they are the broadest. But there are other schemes by developed countries such as Canada and Japan, and lately advanced developing countries such as China have provided preferential market access for selected African exports. However, the rules of origin governing these schemes are not universal. Besides, the requirements on local content tend to be high in most cases, making it difficult for African countries to exploit the preferences optimally. In addition to the high local content requirements, most of the rules of origin do not allow for third country sourcing of raw materials (even where they are cheaper). These restrictive rules of origin therefore pose a major constraint to Africa's ability to fully utilise the preferences and also to deepen the diversification of its economies.

All these challenges, relate to exogenous policy decisions by African trading partners. The current Doha Round has a mandate to address most of these elements. In deed some progress has been made in some of them, and Africa has been very active in the definition of the principles (framework) and now the modalities of how the multilateral rules could be structured in a way that the international market environment is improved.

Yet, even if these externally imposed challenges were to be addressed either through the Doha Round, the EPA negotiations, and other bilateral arrangements, Africa is unlikely to witness any dramatic change in its development prospects through trade for several reasons. In deed, in the context of negotiations, research at the ECA has shown that while there are gains still to be realised from the market access agenda in the WTO, most of these gains could accrue to other developing countries. There will be little benefits for Africa from the multilateral negotiations given the current structure of the African economies, and so their marginalisation in the global trading system is likely to continue. The African countries lack the capacity to exploit the trading opportunities that might be presented by the solutions to the external challenges. And it is this lack of capacity to exploit trading opportunities that AfT is expected to play a major role if effectively operationalised. As will be indicated below, the capacity-enhancing investments in economic and social infrastructure and removal of supply constraints all require significant amount of resources.

Further research within the ECA has also shown that Africa's economies could face the risks of de-industrialisation, for the few African countries that have made some steps towards development of some industries. AfT would be crucial in safeguarding and deepening these gains towards industrialisation, however limited they might have been in the past. Another clear conclusion that studies at the ECA has arrived at is that while there are some benefits from preferences, these benefits could be greater, with the strengthening of internal supply capacities. The AfT would play a major part in such endeavours.

10.5. THE INTERNAL TRADE CHALLENGES

The internal trade challenges that Africa faces are already well documented, and as noted they were indeed the motivation for the need to strengthen development dimensions of the multilateral trading system. The challenges range from institutional to infrastructure deficiencies and this section revisits briefly the main elements.

Weak supply capacities: An important challenge that African countries face universally, are the weak supply capacities. There have not been sufficient investments to build requisite capacities to meet the supply needs of demanding markets. Thus, even for the few exports that African countries are able to produce, lack sufficient production infrastructure means that suppliers from Africa are not always able to meet their demand. The low supply capacities are worsened by poor infrastructure and unreliable business support utilities.

Lack of skills towards better quality products: The markets for different products produced under the different segments of the global supply chain are becoming intensely contested. In particular, other developing countries are gaining ground in these markets, and are able to exploit potentials of vertical integration. African countries on the other hand, due to lack of skills are not able to exploit these possibilities as the products fail to meet the required quality standards.

Limited resources for research and development: In addition to lack of skills, African countries face the challenge of having no or limited resources that would support research and development to aid in development and adoption of better technologies essential for manufactured products diversification. As the *Economic Report on Africa 2007*, argues, research and development is critical to countries' ability to diversify—vertically and horizontally.

Trade finance: Even where African producers are able to secure international markets, an important constraint that they have to overcome relates to trade financing. This includes both pre-financing and insurance. This is an area that is affected by the shallow financial markets and as such most African countries lack instruments that would facilitate trade financing.

Trade facilitation: One of the major internal challenges that Africa has with respect to trade relates to trade facilitation. As noted above, improved market access cannot be fully exploited if means to facilitate trade are weak. Therefore, as African countries have argued repeatedly in the Doha Round, the question of trade facilitation for Africa goes beyond the customs administration issues. These are the software elements of the trade facilitation. Rather, trade facilitation concerns for Africa also involve the weak transportation infrastructure to and from the ports of entry of exports and imports.

Weak port and transportation infrastructure: Africa is probably most challenged in the area of trade by the dilapidated hardware elements of trade facilitation. Beyond the ports of entry and exit of traded goods, many countries lack appropriate national and even connecting infrastructure to their neighbours. Thus, the time taken to reach the market by African products is magnified by the weak infrastructure making it difficult for Africa to trade.

Internal non-tariff barriers: Besides the infrastructure problems, there are intra-African trade barriers that are still a major challenge in the continent. In this regard, Africa is faced with the challenge of removing the intra-African trade barriers, which today make it difficult for the deepening of African integration. The removal of these barriers, when supported with improvements in the trade facilitation mechanisms (to the border, at the border, and after the border) will allow the continent to tackle the small markets problems that make it difficult for African producers and exporters to exploit the economies of scale, which would have the spin-off of helping build competitiveness.

Institutional constraints: The ECA's diagnosis of the failure for African countries to fully exploit their trade potential led to the thinking that besides low productive capacities and poor infrastructures, trade development in African countries—and in particular LDCs—has been hampered by weak trade policy capacities. This lack of appropriate capacities to analyse trade policy issues in African countries led to outcomes that did not contribute to better role for trade in the regions development. The trade capacity constraints in African countries identified in ECA work include:

- Lack of the capacity to negotiate effectively on trade issues of interest to Africa.
- Lack of capacity to influence and set the agenda or pace of multilateral negotiations.
- Lack of capacity to fulfil commitments to the multilateral trading system, without undermining African development goals.
- Lack of the capacity to formulate effective trade policies.
- Lack of capacity to exploit trading opportunities.

10.6. THE RATIONALE OF THE AID FOR TRADE

The foregoing discussion shows clearly that African countries are already at a disadvantaged position, and whatever global trade reforms are to be undertaken; they are unlikely to reap proportionate benefits from trade. It is this realisation that developing countries especially those in Africa are unlikely to make in-roads in the rapidly globalising world that underpinned the initiative by World Trade Organisation (WTO) Members to develop an Aid for Trade package. AfT was officially put on the WTO agenda at the 6th Ministerial conference in Hong Kong in December 2005.

It is worth noting however that the idea of the AfT started gaining currency at the 5th WTO Ministerial Conference in Cancun when the UN Secretary General noted the need for AfT to complement other aid for development. The idea was further deepened in many UN reports and gatherings where it was agreed that there was need to go beyond the soft trade-related technical assistance to the support that would help deepen productive capacities, finance trade-related hard infrastructure, while at the same time deal with the adjustment costs of trade reforms.

This integration of the AfT in the Doha Round was a recognition that there was need to go beyond the traditional trade-related technical assistance. This shift towards acceptance of the importance of the hardware elements of the trade-related support is evident in the mandate given by the WTO member states. The mandate with respect to AfT is "to help developing countries particularly Least-Developing Countries (LDCs), to build the supply-capacity and trade-related infrastructure that would assist them to implement and benefit from WTO agreements and more broadly to expand their trade"[2].

It is also worth recalling that, in spite of the slow progress so far, the need to strengthen trade capacities in developing countries is a major element in the Doha Round. The rationale for the AfT is therefore built around this strong need. But at the same time, it has been accepted that the AfT cannot be a substitute for the development benefits that are aimed for from a successful conclusion of the Doha Round. The Hong Kong Ministerial Decision instructed the Director General (DG) of the WTO to set up the task force that came up with the recommendations on how to implement the AfT. This report was submitted on 27 July 2006 and is the basis of the on-going activities on the AfT initiative.

10.7. OBJECTIVES AND GUIDING PRINCIPLES

In the context of Africa, the objectives of the AfT promise to be an important complement to other forms of aid for development that the region is already receiving. The initiatives objectives includes: enabling developing countries use trade more effectively; helping countries build supply-side capacity and trade-related infrastructure to access markets and export more; helping facilitate, implement and adjust to trade reform and liberalisation; assisting regional integration; assisting smooth integration into the world trading system; and assisting in the implementation of trade agreements. So far, trade-related technical assistance has failed to fully catalyse trade as an engine for growth. This in spite of Africa as a region being more open to international trade than any other developing region except East Asia and the Pacific. Despite its trade openness, the region, as already noted has witnessed a substantial decline in the share of exports relative to the world total. This decline seems even more worrisome, given a marked rise in the relative importance of developing countries as a whole in world exports. Africa could therefore greatly benefit from the AfT were the initiatives objectives to be met.

The Aid for Trade operationalisation will build on existing trade-related assistance mechanism, for example the Integrated Framework (IF) and Joint Integrated Technical Assistance Programme (JITAP). In addition, it will use existing guidelines for aid delivery as outlined in the Paris Declaration on Aid Effectiveness within the five broad areas[3] of ownership, alignment, harmonisation, managing for results and mutual accountability. By adhering to these principles, it is expected that the operation elements of the AfT will result in strengthened demand side as developing countries fully mainstream trade in their national development strategies, leading to strong country ownership of aid programmes and country-based formulation of trade related needs and priorities. On the donors' part, there could be strengthened response to trade-related needs and priorities in terms of policies and coordination. These two results could then lead to a stronger bridge between country demands and donor responses at the country level, regional and global level. Lastly, a strong monitoring and evaluation mechanism is possible if the AfT operationalisation were to be guided by the Paris Declaration principles.

10.8. THE SCOPE OF AID FOR TRADE

To what extent is the AfT likely to address Africa's trade challenges? An examination of the scope of the AfT indicates that there is correspondence with the challenges that Africa faces in trade. This is particularly with respect to the internal challenges identified above and which African countries expected to see in the elements of the AfT (see Box 10.1.). There are six broad categories defining the scope of the AfT.

Firstly, it is expected to deal with constraints in developing countries related to trade policy and regulations. This includes among other things training officials, helping governments implement trade agreements, and complying with rules and standards. The traditional trade-related capacity building has to a large extent been addressing some of these issues.

Second, AfT is also expected to be geared towards trade development that includes providing support services for business, promoting finance and investment, conducting market analysis and e-commerce. Like in the case of support to trade policy and regulations, the focus by AfT on trade development is not new. Traditional technical assistance has been dealing with this. However, the support in this area has not been enough to overcome the market access barriers externally imposed for African exports.

Third, AfT is expected to help in the development of trade-related infrastructure such as building roads and ports. This is a more deepened support for trade assistance, an innovation of the AfT to complement the normal aid for development. At this juncture, it is agreed that there will be additional funds for this component of the AfT, and it will not be a reallocation of already committed development assistance.

Box 10.1. Key elements for AFT identified by African countries

- Enhance productive capacities of African countries, through increased productivity, addressing the necessary regulatory reforms, human resource and physical infrastructure needed to make African economies competitive;
- Support investment in improving competitiveness of African countries
- Assist African countries to invest in removing impediments to business that drive up costs for exporters;
- Assist African countries to invest in infrastructure;
- Help African countries to effectively deal with adjustment costs associated with trade liberalisation, such as rising food prices for net-food importing countries, preference erosion, reduction in tariff revenues, and other economic and social costs;
- Assist African countries to integrate trade policy in their overall development strategies and create a strong public-private sector partnership; and
- Help African countries to invest in human resource development
- Support to regional integration initiatives

Fourth, AfT is expected to provide support for building productive capacity, which comprises of improving the capacity of a country to produce goods and services. This will enable countries to deepen their diversification through development of new products. AfT will therefore facilitate value chain analysis, and then will support the building of capacities for the development of products that help African countries exploit those segments of the value chain where they have comparative advantage.

Lastly, many studies have clearly shown that implementation of trade reforms is not costless. The ECA work on the impacts of different modalities options for agriculture and Non-Agriculture Market Access (NAMA) have shown that beyond the economic costs, are social implications associated with de-industrialisation and preference erosion. It is anticipated that the AfT will provide financial assistance to enable developing countries meet adjustment costs associated with trade policy reform, including balance of payment problems resulting from lost tariff revenues or from erosion of preferential market access. At this point, it helps to note that the AfT is also expected to help developing countries implement other trade agreements. And while the Doha Round commitments will present adjustment challenges, the same could be said with respect to Africa's commitments under regional integration agreements and the Economic Partnership Agreements (EPAs) under negotiations with the EU. The AfT scope could allow it to be used in support of the implementation of trade aspects of the EPAs[4].

10.9. ADDITIONALITY OF RESOURCES UNDER THE AID FOR TRADE

Even before the exact estimates of the AfT requirements, an important question as noted already is whether AfT will deliver "new and additional resources" to support technical assistance programmes and projects. This has been an important issue for African countries. Consequently, it will be important for the AfT not to be a "re-packaging" of existing technical assistance and projects, for this would not serve African countries address the challenges hindering them from benefiting from the global trade. As such, the AfT will need to be adequately funded, with the additional resources predictable and sustainable as well. African countries position has been very clear on the question of additionality of resources (see Box 10.2. on African countries position on financing the AfT). The additionality of resource flows must hinge on other specific issues as identified in the AfT. These include the scope of AfT that will define the resources needed; the content and elements of the initiative, which will define the programmes and projects to be supported; mechanisms for mobilising the resources; modalities for implementation; and system of monitoring and evaluation.

10.10. AFRICA REGIONAL REVIEW ON AID FOR TRADE

In this section, the chapter presents the results of Africa's regional review meeting in Dar-Es-Salaam, Tanzania held on 1 and 2 October 2007. These results feed into the global review meeting in Geneva in November 2007. The overall goal of the review meeting was the need to move the AfT initiative from the Task Force Recommendations to action. The following were the key objectives of the meeting:

- Underline trade's central importance to the achievement of Africa's development goals;
- Identify the key priorities for strengthening capacity and increasing trade;
- Emphasise the importance of comprehensive trade strategies (regional as well as national);
- Highlight the need for increased and effective financing; and
- Secure political commitment on the way forward.

10.10.1 Key Issues from the African Regional Review of the Aid for Trade

The African review meeting was organised jointly by the ECA, WTO and ADB and it brought together African Ministers of Finance, Ministers of Trade, bilateral and multilateral donors and the private sector. Several key points can be summarised from the discussions that took place between the African policymakers, the private sector, regional development banks and the international partners.

First, there is now consensus that that AfT matters for Africa. This consensus is underpinned by the realisation that African products are competitive at the factory and farm-gate level when compared with those of Asian countries. However, the African products lose their competitiveness when it comes to placing them in the international markets. The African review meeting concluded that for the continent to be able to compete and succeed in the global market place, the AfT is necessary, as it would help the countries to diversify and address the cost of production factors. The regional dimension of the AfT programmes would however be critical as this would both help African countries optimise their regional integration processes, but at the same time exploit economies of scale through larger competitiveness-building markets. For this to happen, new instruments by the development banks to finance regional competitiveness-enhancing strategies will have to be developed. And for this to be successful, the competitiveness-enhancing strategies must at the same time address specific needs of each country and must have the private sector as the key anchor.

Second, it was clear from presentations by the African RECs and interventions from the African Ministers that important initiatives at national and regional level aimed at enhancing competitiveness exist. Many of these projects and programmes already identified have been geared towards strengthening of production capacities for trade. A key issue that the review meeting identified was the need to resolve the lack of coordination in the existing strategies. In particular, bilateral and multilateral development partners have separate and uncoordinated programmes with individual countries and regional groupings, and this would have to be addressed to enable seamless super-imposition of the AfT on these existing strategies.

Third, the African review noted the broad array of areas where AfT support is required especially among the RECs, which then calls for clear prioritisation. But most significantly, it was evident that the RECs priorities narrow for now to just three key priority areas, albeit with different ranking for each REC. Moreover, the existing strategies are at different stages of implementation and financing gaps are a major constraint. The African review meeting therefore concluded that it was important that the concrete implementation of the AfT for Africa focus on the following: *infrastructure, trade facilitation,* and *standards.* AfT resources that are targeted at the three areas would help address the paradox of African products, which are competitive at the factory and farm-gate, but uncompetitive at the global markets.

Fourth, the success in implementation of competitiveness-enhancing strategies is contingent upon the AfT also strengthening the human capital development in the continent. This should go hand-in-hand with the deepening of the private sector involvement. Only then can the AfT provide the catalytic effect for economic change as if fills the financing gaps of existing and new strategic areas.

10.10.2 Conclusions from the Africa Aid for Trade Review Meeting

Leadership the key to success: AfT initiative will have a better chance of succeeding if there is leadership at the national level including on the choices that countries have to make. A pertinent aspect of this leadership relates to clear political decisions that underpin trade and development as an idea upon which countries' future could be built upon. A secondary aspect is the need to build the agenda for trade and development on strengthening national and regional capacity once the political choice is made, as and this must be reflected in the national development plans.

Delivery mechanisms: Existing mechanisms were favoured when it came to the point of implementation. The desire for new mechanisms could only end up duplicating existing structures. It was however noted that the parallel Enhanced Integrated Framework (EIF) would continue to serve the LDCs.

Some of the African priorities: A priority that came up which the AfT must immediately tackle relates to the *infrastructure constraints* and also the maintenance of the infrastructure. The other important obstacle to private sector competitiveness is the *trade facilitation* issue including at the border that AfT must target. In both of the above priorities, the regional integration, hence regional dimension is a crucial factor towards competitiveness building. Therefore, *capacity for the RECs* emerged as a priority area for Africa's AfT. A fourth priority, which is part and parcel of the infrastructure and trade facilitation strategies, was the *involvement and development of the private secto*r. In fact, the private sector is on the one hand a player from demand side but also an actor in the supply of AfT. Furthermore, for Africa, in the context of the private sector, the SMEs were identified, as a crucial component/segment given that it is the SMEs that suffer most with the indirect and direct trading costs, and as a result AfT measures must aim to strengthen their competitiveness. Finally, public-private partnerships hold important leverage for the success of the AfT in Africa.

10.11. LOOKING AHEAD: MOBILISING AID FOR TRADE IN AFRICA

As indicated above, the African countries met in Dar-Es-Salaam, Tanzania on 1 and 2 October 2007 to discuss the best ways to mobilise the political and financial commitments required to realise the objectives and goals of the AfT. But beyond the discussions in Dar-Es-Salaam, there are important issues that the African countries and their partners will have to address in order to achieve effective results from implementation of the recommendations from the Dar-Es-Salaam meeting. These issues are areas of work in themselves for the African countries and the donors as well and they include:

- Mainstreaming of trade in the development strategies of African countries.
- Aligning Donor Strategies with country and regional needs and the global partnership goals as internationally agreed.
- Translating the diagnostics into actionable programmes in which case the regional needs assessments are developed into activities that can be implemented with clear focus on expected outcomes.
- Improving on the delivery mechanisms. In particular, the African countries will have to finalise the regional instruments e.g. the regional funds in which priorities are as agreed among the member states. The effectiveness of the regional delivery mechanisms will depend on actions at national level given that the AfT programmes will cut across national borders.
- Developing monitoring and evaluation frameworks for the AfT programmes.
- Ensuring the participation of the private sector including having the right incentives that would justify its participation especially in regional projects.
- Coordinated support from the donor community. The lessons learnt from IF and JITAP and the commitments under Paris Declaration should guide the response on this issue of donor coordination.

10.12. CONCLUSION

This brief chapter has discussed the challenges that Africa faces in the area of trade and how the scope of the Aid-for-Trade is likely to address these challenges. Operationalising of the AfT initiative should come up with national and regional mechanisms that would effectively address these trade-related challenges as highlighted in the paper in looking ahead. The chapter has also shown that there is consensus on the importance of AfT operationalisation in Africa, judging by the outcomes of the African Regional Review meeting on the initiative. The chapter also suggests some of the issues that will have to be addressed as the international community and the African countries look ahead to the mobilisation of AfT resources for Africa.

ANNEX: RECOMMENDATIONS OF THE TASK FORCE ON AID FOR TRADE[5]

Mandate

The Hong Kong Ministerial Declaration invited the WTO Director-General to create a Task Force to provide recommendations "on how to operationalise Aid for Trade" and "on how Aid for Trade might contribute most effectively to the development dimension of the DDA". It states that "Aid for Trade should aim to help developing countries, particularly LDCs, to build the supply-side capacity and trade-related infrastructure that they need to assist them to implement and benefit from WTO Agreements and more broadly to expand their trade. Aid for Trade cannot be a substitute for the development benefits that will result from a successful conclusion to the DDA, particularly on market access."

Rationale

Aid for Trade is about assisting developing countries to increase exports of goods and services, to integrate into the multilateral trading system, and to benefit from liberalised trade and increased market access. Effective Aid for Trade will enhance growth prospects and reduce poverty in developing countries, as well as complement multilateral trade reforms and distribute the global benefits more equitably across and within developing countries.

Financing

Additional, predictable, sustainable and effective financing is fundamental for fulfilling the Aid-for-Trade mandate. The effectiveness of the following recommendations for operationalizing Aid for Trade requires substantial additional targeted resources for trade-related programmes and projects as pledged at the WTO's Hong Kong Ministerial Conference, and against the background of the broader international commitment at the UN's Monterrey Conference and the G8 Summits in Gleneagles and St. Petersburg to significantly scale up development assistance by 2010. The Task Force urges the Director-General to seek confirmation from donors and agencies that funds are readily available for the implementation of the Aid-for-Trade initiative as part of his mandate to consult on "appropriate mechanisms to secure additional financial resources for Aid for Trade".[6] In order to measure additionality and the adequacy of funding available to meet the Aid-for-Trade needs of developing countries, including those associated with a successful completion of the DDA, an account of what is being done today needs to be established as part of that process. The Task Force urges donors and agencies to provide the necessary information in order to make it possible for the Director-General to fulfil his mandate.

Scope

The scope of Aid for Trade should be defined in a way that is both broad enough to reflect the diverse trade needs identified by countries, and clear enough to establish a border between Aid for Trade and other development assistance of which it is a part. Projects and programmes should be considered as Aid for Trade if these activities have been identified as trade-related development priorities in the recipient country's national development strategies. In this regard, it should be pointed out that while the PRSPs[7] reflect national development priorities for some countries, other development strategies are equally important and will need Aid-for-Trade financing. At the same time, clear and agreed benchmarks are necessary for reliable global monitoring of Aid-for-Trade efforts to assure accurate accounting and to assess additionality. The following categories, building upon the definitions used in the Joint WTO/OECD Database, have been identified:

(a) *Trade policy and regulations, including:* Training of trade officials, analysis of proposals and positions and their impact, support for national stakeholders to articulate commercial interest and identify trade-offs, dispute issues, institutional and technical support to facilitate implementation of trade agreements and to adapt to and comply with rules and standards.
(b) *Trade development, including:* Investment promotion, analysis and institutional support for trade in services, business support services and institutions, public-private sector networking, e-commerce, trade finance, trade promotion, market analysis and development.
(c) *Trade-related infrastructure, including:* Physical infrastructure
(d) *Building productive capacity*
(e) *Trade-related adjustment, including*: Supporting developing countries to put in place accompanying measures that assist them to benefit from liberalised trade.
(f) *Other trade-related needs*

Reporting on categories (a) and (b) should follow the definitions in the Joint WTO/OECD Database. The activities that fall outside of the current Joint WTO/OECD Trade Capacity Building Database definition, i.e. category (c), (d) (e) and (f) should be reported as Aid for Trade when these activities have been explicitly identified as trade-related priorities in the recipient country's national development strategies, such as the PRSP.

Challenges/Gaps

Since the start of the DDA in 2001, donors have stepped up their commitments on trade-related assistance. More developing countries are also integrating trade into their development strategies. But major challenges remain. These can include:

- Low attention to trade as a tool of development in recipient countries and in donor agencies.
- Insufficient trade mainstreaming in national development strategies and PRSPs.
- Lack of private-sector involvement in identifying trade needs.
- Limited absorptive capacity in recipient countries.
- Inadequate linking mechanisms and lack of predictability in donor response to trade priorities identified at the national and regional levels.
- Lack of coordination and coherence in donors' trade-related response.
- Slow, duplicative and bureaucratic processes in the assessment and delivery of trade assistance, including burdensome parallel structures within recipient countries.
- Lack of data on, and analysis of, trade polices and their impact on development, lack of easily-available information on existing Aid-for-Trade instruments.
- Ineffective monitoring of trade-related country policies and donor activities; absence of rigorous, independent project and programme evaluation and impact assessment.
- Limited support for regional, sub-regional and cross-border trade-related programmes and projects.
- Inadequate support to address the adjustment costs of trade liberalisation.
- Insufficient resources for infrastructure and productive capacity building.
- Uneven country coverage.

Operationalizing Aid for Trade

(a) Objectives

- To enable developing countries, particularly LDCs, to use trade more effectively to promote growth, development and poverty reduction and to achieve their development objectives, including the Millennium Development Goals (MDGs).
- To help developing countries, particularly LDCs, to build supply-side capacity and trade-related infrastructure in order to facilitate their access to markets and to export more.
- To help facilitate, implement, and adjust to trade reform and liberalisation.
- To assist regional integration.
- To assist smooth integration into the world trading system
- To assist in implementation of trade agreements.

(b) Guiding principles

Aid for Trade should be guided by the Paris Declaration on Aid Effectiveness, applicable to all parties involved (donors, agencies and beneficiaries), including key principles such as country ownership, mutual accountability, aligning aid to national development strategies, effective donor coordination, harmonisation of donor procedures, use of programme-based aid modalities, managing for result, transparency, and predictable and multi-year commitments, which should be built into all programming. Aid for Trade should be rendered in a coherent manner taking full account, *inter alia*, of the gender perspective and of the overall goal of sustainable development. Administrative costs associated with the delivery of Aid for Trade should be minimised to ensure that the resources go to the actual implementation of identified priority projects and programmes. The competence and skills of the human resources available at national and regional levels should be used in an optimal way.

(c) Strengthening the "demand side"

A commitment to country ownership and country-driven approaches – as well as a commitment of governments to fully mainstream trade into their development strategies – is key to the effectiveness of Aid for Trade. In some countries, the processes for mainstreaming trade into national development strategies, for formulating trade strategies, and for proposing priority trade projects for donor financing, need to be strengthened through technical assistance and capacity building to help developing countries put in place effective and sustainable trade policy frameworks and processes. Where consultative mechanisms already exist, they can be used – or improved upon. Value-chain analysis could be one valuable tool to identify trade needs.

The Enhanced Integrated Framework (IF) for LDCs. The purpose of the IF is to strengthen the LDCs' trade capacity, including the ability to identify their trade needs and to propose priorities to be supported by development partners. The recommendations on an enhanced IF, as agreed by the Integrated Framework Steering Committee (IFSC), will be an essential foundation for strengthening the demand-side of Aid for Trade in LDCs.

Non-LDCs. Many other developing countries also need support to mainstream trade into national strategies, to establish broad-based consultation processes involving the private sector, civil society organisations and relevant government agencies to formulate trade strategies, to develop action matrices, and to formulate priority project proposals.

Regional needs. Some of the constraints facing developing countries are regional, sub-regional or cross-border in nature. These needs should be identified and properly addressed. Regional organisations, including regional banks, regional integration organisations and regional economic communities, may play a role in assisting countries to identify such needs.

Recommendations

- Implement the recommendations for an enhanced Integrated Framework.
- Establish effective national coordination, involving all relevant stakeholders, including the private sector, with a view to identifying the strengths and weaknesses of economies as a whole, and the particular challenges facing the trade sector.

- Explore the necessity of establishing a similar, but separately funded, in-country-process for non-LDCs "International Development Assistance (IDA)-only" countries, if such mechanisms do not already exist or can be improved upon.
- Urge agencies, donors and governments in other developing countries to work together to establish similar processes if they do not already exist. These processes should be modelled to the specific circumstances and needs of the country concerned, building on what already exists where possible and appropriate.
- Urge donors and agencies, together with regional banks and organisations, to step up their efforts to identify regional, sub-regional and cross-border needs, including those related to regional integration.
- Establish a system of data collection and analysis at country level.

(d) Strengthening donor "response"

Donor policies. Donors should give more attention to trade issues in their aid programming and strengthen their trade expertise both in the field and at headquarters. There is a need for improved coordination of staff working across sectors and for greater trade mainstreaming in aid agencies' programmes.

Donor coordination. Greater donor and agency coordination and harmonisation of procedures – at both the local and global level – is critical. Trade-related programmes and projects should be more coherent, both in terms of operations and policy.

Donor response. In allocating resources for Aid for Trade, donors and agencies should be guided by priority projects and programmes identified by developing countries, as well as by their potential merit in relation to the objectives for Aid for Trade. These priorities should be mirrored by donor and agency support. Each agency would need to determine how to deploy or reorient its financial and technical assistance to support either capacity building or accompanying measures related to trade liberalisation.

Recommendations

Donors and agencies should:

- integrate trade and growth issues more effectively in their aid programming;
- further strengthen their trade expertise both in the field and in capitals;
- use needs assessment processes (where available), and their results, as a basis for their programming;
- move towards a programme/sector/budget approach, if country owned, if mainstreamed in national development strategies and if a robust system of financial accountability is in place;
- make targeted funds available for building infrastructure and removing supply-side constraints – over and above capacity building and technical assistance – perhaps as co-financing with multilateral development banks; and
- consider channelling Aid-for-Trade Funds multilaterally, when appropriate.

(e) Strengthening the bridge between "demand" and "response"
Country level

Matching. Strengthened in-country structures, with improved links to donor financing, are needed to help move from trade-related diagnostics to implementation, and to maximise access to multilateral and bilateral resources. The task of matching demand for Aid-for-Trade projects with response could be addressed by strengthening national coordination through a "National Aid-for- Trade Committee", which would include recipient countries, donors, and other relevant stakeholders, such as the private sector, under the leadership of relevant ministries. This committee should complement – not replace – existing PRSPs and other coordination mechanisms. If needed, this process could be supported by agencies that could serve as a clearing house.

Mainstreaming trade. Effectiveness in implementing Aid for Trade will depend on many actors working together in a coherent way. It will involve, for example, the World Bank, the IMF, regional development banks, UN agencies and donors at the national as well as the international level, and trade, agriculture, development and finance ministries at the national level. It is the responsibility of donors, agencies and recipients to do their part in reforming how those entities integrate trade into development and national strategies.

South-South cooperation. Technical cooperation among developing countries is a valuable tool to deliver effective results because of their common experience and understanding of the challenges they face. The valuable technical expertise of the South could be used to implement projects through triangular schemes of cooperation.

Private sector. As actors in the field, private enterprises are well placed to identify trade-related problems and bottlenecks. An increased dialogue between the public sector and private entrepreneurs would improve effectiveness in assessing Aid-for-Trade needs, in diagnostics, and in implementation, as well as in evaluating effectiveness in implementation.

Recommendations

- Recipient countries should mainstream trade into national strategies, such as PRSPs, formulate trade strategies, and propose priority trade projects for donor financing.
- The division of responsibility for funding and implementing Aid-for-Trade projects and programmes should be addressed through country-based processes such as PRSPs or Consultative Groups, if necessary complemented with a partner conference focusing specifically on trade-related support, convened once countries have integrated trade into their national strategies.
- A National Aid-for-Trade Committee could be established, where necessary, to ensure trade mainstreaming in national development strategies, determine country needs, set priorities, assist in matching "demand" and "response", and help in

evaluation. Tasks could include identifying co-financing or leveraging funds from other larger funds, as well as assessing adjustment needs and brokering financing for such programmes. Recipient countries could request agencies to perform a coordinating role.

- Partners should commit to contributing to the implementation of trade strategies and identified priority projects and programmes. The resulting plan should incorporate a results-based management framework resting on – and reinforcing – mutual accountability. Indicators of progress should be agreed.
- Promote the involvement of local, regional and private-sector actors, as well as South-South cooperation through triangular schemes.

Regional level

Many countries require cross-border infrastructure and regional policy cooperation to trade more effectively. The ability to identify cross-border and regional needs should be strengthened at the country, regional and multilateral level. Once needs have been identified, donors and agencies must improve their ability to respond. In particular, assistance in formulating and financing accompanying measures could help to make regional integration an effective building block for the multilateral trading system. At the forthcoming September Development Committee Meeting, strengthening support for regional, sub-regional and cross-border needs will be discussed.

Recommendations

- Strengthen the following functions in relation to regional, sub-regional and cross-border issues:
  - diagnosis of needs;
  - costing of projects;
  - preparation of project proposals; and
  - the coordination of donor response, including brokering and co-financing of needs that at present are difficult to finance through country-based processes, (e.g., cross-border infrastructure and policy-integration projects).
- Assign responsibility for these functions. In doing so, priority should be given to improving and strengthening existing mechanisms, including those at the multilateral and regional level, before considering a new mechanism. In exploring the most efficient solution, the conclusions from the discussions at the forthcoming Development Committee should be taken into account. Any solution should involve all relevant stakeholders and give priority to existing regional integration programmes that lack funding.
- Explore the merits of establishing a Regional Aid-for-Trade Committee, comprising sub-regional and regional organisations and financial institutions, to oversee the

implementation of the sub-regional and regional dimensions of Aid for Trade, to report on needs, responses and impacts, and to oversee monitoring and evaluation.

Global level

A number of tasks in relation to Aid for Trade are best performed at the global level. These include:

Data collection. Lack of empirical data has made it difficult to examine the relationship between policies related to trade and development performance. Better data and statistics are a precondition for better understanding the process of globalisation and its impact, and for determining priorities for development cooperation.

Knowledge creation and sharing. Dissemination of Aid-for-Trade evaluation results, development of best practices and guidelines, and facilitation of information sharing, involving all relevant actors, needs to be improved at the global level, in order to assure efficient use of Aid-for-Trade funds.

Channelling donor funding. Some donors might wish to direct Aid-for Trade funds through multilateral channels, which would allow them to support Aid for Trade without having to build their own institutional capacity in this area and without getting involved at country level. This could include providing support for processes similar to the IF for non-LDC IDA-only countries.

Matching. While a clearing-house function should in most cases be performed at the country and the regional level, sessions dedicated to specific themes and groups of countries could be periodically organised to provide a platform for donors and developing countries to discuss specific gaps which may occur in the implementation of Aid for Trade. One important function could be to connect outstanding Trade-Related Assistance (TRA) needs to donors willing to contribute to their fulfilment.

Recommendations

- Strengthen the following functions in relation to global issues:
  - the collection and analysis of data on trade policies and their impact, the facilitation of knowledge sharing, and the development of guidelines. Funding for such activities needs to be secured;
  - provision of information on existing Aid-for-Trade instruments and expertise; and
  - matching and brokering unfunded TRA-needs and available donor funding for such projects and programmes.
- Assign responsibility for these functions. In doing so, priority should be given to improving and strengthening existing mechanisms before considering the establishment of a new clearing house at the global level.

(f) Strengthening monitoring and evaluation

Monitoring and evaluating progress is essential in building confidence that increased Aid for Trade will be delivered and effectively used. It will also provide strong incentives to both donors and recipients to advance the Aid-for-Trade agenda. It is important to emphasise the need for concrete and visible results on the ground. All the providers of Aid for Trade and the recipient countries have the responsibility to report on progress and results.

Monitoring. In recipient countries, monitoring should cover trade mainstreaming in national strategies, such as PRSPs, the identification of priority needs, donor responses, progress in implementing trade-related projects and programmes as well as the impact of these efforts. Donors who have made commitments to Aid for Trade should report on the content of such commitments as well as on how they plan to meet the targets for Aid for Trade that they have announced.

Evaluation. Rigorous Aid-for-Trade programme evaluation is particularly important because projected significant increases in Aid for Trade may stretch the delivery capacity of donors and the absorptive capacity of recipients. In-depth country-impact evaluations of Aid for Trade programmes should be undertaken to build knowledge and facilitate a results-based approach to delivery. Evaluation of in-country processes should focus, *inter alia*, on progress in mainstreaming trade in national development plans. Evaluations should adopt a results-based approach in order to ensure effectiveness of Aid-for-Trade programmes in relation to the objectives.

Recommendations

- A global periodic review of Aid for Trade should be convened by a monitoring body in the WTO, based on reports from several different sources, to be published if feasible on the WTO web page:
  - from the country level;
  - from donors;
  - from the regional level;
  - from relevant multilateral agencies; and
  - from the private sector.
- Mechanisms to facilitate reporting to the global monitoring body should be enhanced, including the possibility of a notification process for WTO Members.
- The global periodic reviews should be followed by an annual debate on Aid for Trade convened in the WTO General Council to give political guidance on Aid for Trade.
- Recipient countries should report on the trade mainstreaming in national development strategies, such as the PRSPs, the formulation of trade strategies, Aid-for-Trade needs, donor responses, and implementation and impact. The primary

responsibility for reporting to the global monitoring body would lie with the National Aid-for-Trade Committee.

- Donors should report on funds dedicated for Aid for Trade, how they intend to meet their announced Aid-for-Trade targets, the Aid-for-Trade categories covered, and their progress in mainstreaming trade into their aid programming.
- Multilateral and regional actors should be encouraged to report regularly on their Aid-for-Trade activities, progress and impact. When appropriate these actors – including the OECD/DAC – should be asked to assist in providing input and in the organisation of the periodic Aid-for-Trade review in the WTO.
- The private-sector should be provided an opportunity to report on their Aid for Trade contributions.
- An assessment of Aid for Trade – either as a donor or as a recipient – should be included in the WTO Trade Policy Reviews.
- Evaluation of country-needs identification, trade mainstreaming in national strategies and PRSPs, donor response and impact on the ground in relation to stated objectives, should be promoted and funded.
- The scope of the Joint WTO/OECD Database should be reviewed in light of the Task Force's definition of Aid for Trade. It should also be updated based on more accurate identification of needs (and the responses) by both providers and recipients of Aid for Trade.

How Aid for Trade can contribute to the Development Dimension of the DOHA-Round

Aid for Trade is important in its own right. It should assist developing countries to benefit from increased trade opportunities multilaterally (both from previous rounds and from the anticipated results of the DDA), regionally, bilaterally and unilaterally. The Task Force therefore recommends that Aid for Trade must be operationalised as soon as possible. At the same time, the Task Force affirms that Aid for Trade is a complement, not a substitute, for a successful Doha Round. Increasing trade opportunities for developing countries, in particular the least-developed among them, remains the most important contribution that the WTO can make to development. A successful conclusion of the Round will increase the need for assistance to implement new agreements (e.g., Trade Facilitation), to ease adjustment costs, and to make use of new market access. Aid for Trade is a complement to the Doha Round, but it is not conditional upon its success.

Next Steps

These recommendations are directed to many different actors. The Task Force suggests the following next steps:

- urges Members to expeditiously implement the recommendations of the Task Force.

- urges the Director-General to use these recommendations in pursuing his mandate to consult on "appropriate mechanisms to secure additional financial resources for Aid for Trade" so that the joint mandate in Paragraph 57 of the Hong Kong Declaration can be implemented in a holistic manner.
- invites the Director-General to communicate these recommendations to relevant agencies and organisations and to urge Ministers at the upcoming Development Committee Meeting in Singapore to give consideration to these recommendations and to encourage the Bank and the Fund to ensure adequate follow-up and to report on the results at the 2007 Annual meeting.
- invites the Director-General to continue, under his coherence mandate, a dialogue on how recommendations targeted at the agencies could be implemented, including where responsibility for implementation should lie.
- invites the Director-General to establish an ad hoc consultative group to take forward the practical follow-up of these recommendations.
- invites the Director-General to begin examining how to implement the recommendations regarding WTO monitoring of Aid for Trade.
- invites the Director-General to convene, at an appropriate time, an initial review of Aid for Trade, with the participation of all relevant stakeholders.
- suggests, after the completion of the DDA, that the Secretariat conduct an assessment of associated Aid-for-Trade needs in developing countries, particularly those most affected, including LDCs, and of how Aid for Trade can contribute to the development dimension of the DDA.

NOTES

1 Regional review meeting on *Mobilising Aid for Trade: Focus Africa* organised by the WTO, jointly with the African Development Bank and the UN Economic Commission for Africa, and hosted by the Government of the United Republic of Tanzania. The meeting which took place on 1–2 October 2007 was the last of three regional meetings organised by the WTO in preparation for the Global Review meeting in November 2007. The other two meetings took place in September in Lima, Peru for Latin American Region and Manila, Philippines for the Asian Region.

2 Paragraph 57 of the Hong Kong Ministerial Declaration., WT/MIN/DEC, 22 December 2005.

3 OECD (2005) Paris Declaration on Aid Effectiveness, DCD/DAC/EFF (2005) Final. OECD Paris.

4 The European Union has already indicated that the large proportion of its Aid for Trade pledge is going to be used for the implementation of the EPAs by the ACP countries.

5 WTO/AFT/1 27 July 2006

6 In Hong Kong, Japan announced development assistance spending on trade, production and distribution infrastructure of $10 billion over three years, the US announced Aid-for-Trade grants of $2.7 billion a year by 2010, and the EU and its member States announced trade-related development assistance spending of €2 billion per year by 2010.

7 Poverty Reduction Strategy Papers (PRSPs) describe the macroeconomic, structural and social policies and programmes that a low income country will pursue over several years to promote broad-based growth and reduce poverty, as well as external financing needs and the associated sources of financing. They are country-led, country-written documents prepared by governments through a participatory process involving domestic stakeholders and external development partners, including the World Bank and the IMF.

ECLAC

CHAPTER 11

AID FOR TRADE: STRENGTHENING LATIN AMERICAN AND CARIBBEAN INTERNATIONAL LINKAGES AND REGIONAL COOPERATION

11.1. INTRODUCTION

The "development dimension" of the Doha Round has meant a clear shift away from the view that trade liberalisation on its own would provide significant benefits for all developing countries(best expressed in the catchphrase "*trade not aid*"), towards a more development-minded approach based on the idea that: (i) many countries need additional, complementary reforms and investments that can be financed by aid (*aid for trade*) if they are to fully exploit trade opportunities; and (ii) certain types of liberalisation may have significant short-term costs for some countries. The rationale for Aid for Trade is therefore that, while trade is considered crucial to their development, these countries continue to face serious challenges in infrastructure and supply-side constraints that are preventing them from taking full advantage of trade liberalisation.

The Aid for Trade (AfT) initiative of the World Trade Organisation (WTO) constitutes an explicit recognition by the international trading community that the global benefits of liberalisation will only be fully reaped if and when developing countries manage to strengthen their productive supply capacity, their institutions and their trade-related infrastructure. In line with the Hong-Kong Declaration,[1] a central contribution of the Task Force created by the WTO in 2006 is not only its mandate to increase the volume of aid but also to improve the management, supervision and control mechanism of AfT flows, by way of active participation of countries (beneficiaries and donors), international organisations and private-sector stakeholders.[2]

The importance of the AfT initiative for Latin America and the Caribbean derives from several interrelated aspects: (i) it has helped to position trade as a key component of national development strategies and highlight the need to mobilise and manage trade-related assistance more effectively; (ii) AfT grants can provide crucial "seed" money for larger infrastructure programmes and other supply–related interventions that often require non-concessional financing; (iii) some countries, which are still eligible for concessional funds, have a significant stake in, and much to gain from, additional funding and more effective implementation of AfT initiatives; and (iv) it has once again underscored the regional dimension of trade and the large gaps in coordinating and financing regional infrastructure and other Regional Public Goods (RPBs) initiatives that are conducive to trade (IDB, 2006, 2007).

P. De Lombaerde, L. Puri (eds.), *Aid for Trade: Global and Regional Perspectives*,
United Nations University Series on Regionalism 2,
DOI 10.1007/978-1-4020-9455-2_12,

Needless to say, the region comprises mostly middle-income countries, though some are also small and vulnerable economies and/or landlocked (and transit) countries. The income status excludes most countries of the region not only from concessional financing from international financial institutions, but also from recent trade-related cooperation initiatives targeted at Least-Developed Countries (LDCs).

Since the early 1990s, regional exports have maintained a marked upward trend, not only compared to developments in other parts of the world but also in relation to the region's own economic history. However, this strong export performance has not, in general, been reflected in the region's rate of GDP growth. Recent regional experience shows that market opening alone is not sufficient for sustained exports and economic growth; strategies that take a narrow view of the key factors that enable trade to contribute to growth have not proven effective in addressing job creation and social equity and in preventing the erosion of competitiveness and stimulating export diversification.

In this respect, there is a general consensus in Latin America and the Caribbean that the instruments of trade policy should be complemented by other public policies and by appropriate institutional mechanisms between public and private actors in order to enable governments, private-sector agents and civil society to fully benefit from the opportunities stemming from international trade. In this regard, the direct and indirect effects of intergovernmental trade agreements – multilateral, bilateral, plurilateral and inter-regional – on the economy as a whole and on the export sector in particular, should be assessed. For exports to function as a cumulative process of learning and technology absorption, both for local trade actors and for the country as a whole, governments can support the private sector by ensuring that international insertion becomes a public good.

For the countries of Latin America and the Caribbean to overcome these obstacles, specific programs should be put in place and sufficient funds allocated for strengthening countries' trade-related technical assistance and capacity building (TRTA/TCB) to encompass the following areas: (i) formulation and implementation of a trade-development strategy and creation of an enabling environment, including improvement of the business environment, to increase the volume and value-added of exports, diversify export products and markets, overcome supply-side constraints and generate positive spillovers from trade to development; (ii) incorporation of knowledge and value-added activities in goods (including natural resources, maquila type operations) and services; (iii) strengthening and/or creating appropriate human and institutional capacity necessary for technological advance and innovation in the export sector; and (iv) enhanced international insertion of the countries of the region in the world economy and greater participation in international value-chains and foreign investment abroad.

Trade capacity building (TCB) for Latin America and the Caribbean has a strong regional dimension: regional integration serves as a vehicle for the enhancement of international competitiveness, export diversification, by product and destination, deepening of foreign direct investment (FDI) and technology upgrading. There is an urgent need to address the issues of how to: (i) stimulate regional trade flows; (ii) reform regional integration institutions to face the challenge of the multiplication of bilateral trade agreements with countries within the region and beyond; (iii) proceed with the convergence of trade-related rules and regulations at subregional levels that is necessary for such trade reactivation; and (iv) improve the systemic competitiveness of individual countries and the region as a whole, by enhancing physical infrastructure (e.g., transport, ports and customs, telecommunications and energy), complementarities and interconnection. Trade-related capacity building measures also include investments, South-South trade expansion and technology transfer.

To address this challenge, the United Nations Economic Commission for Latin America and the Caribbean (ECLAC) is evaluating several options in the framework of its institutional mandate to *contribute to strengthening the international participation and regional integration* of the countries in the region. Given the regional specificities and marked heterogeneity between the countries, this challenge must be addressed with creativity, while at the same time preserving the progress that has been made so far, complementing, not duplicating, existing programs and initiatives, and evaluating the region's specific needs to facilitate ongoing initiatives and/or address others not already considered.

The pillar of AfT proposals from the Economic Commission for Latin America and the Caribbean (ECLAC) is based on establishing programmes and projects, communication and dialogue forums to combine efforts and resources with a view to maximizing aid efficiency on the basis of: (i) management excellence; (ii) stable and sufficient financing; (iii) complementarity and communication between organisations and programmes in charge of AfT; and (iv) creation of adequate conditions and mechanisms for more efficient monitoring and evaluation.

As ECLAC has been pointing out for some years, the present commodity price boom and healthy current-account positions of the region represent an opportunity for Latin America and the Caribbean to lay the foundations for sustained growth in the medium term. With greater public-sector resources and access to low-cost financial markets, the region's countries could enhance their integration into the international economy by increasing infrastructure and human-capital investment, encouraging private-sector entrepreneurial efforts to add value, know-how and innovation to exports and pursuing key initiatives that can contribute to regional integration.

11.2. CURRENT DEBATE: MORE FINANCIAL RESOURCES, MORE EFFICIENCY OR BOTH?

11.2.1 The Trend of Aid for Trade

Between 2001 and 2005, total Official Development Assistance (ODA) rose by 86 per cent.[3] According to estimates from the Organisation for Economic Co-operation and Development (OECD), a similar increase is expected for the 2004–2010 period, which would increase ODA from US$ 87 billion to US$ 130 billion. The ODA

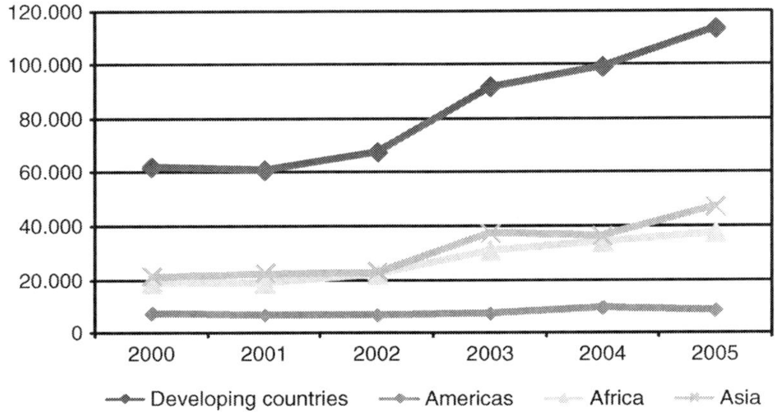

FIGURE 11.1. ODA commitments, 2000–2005 (US$ millions, current prices) (*Source:* OECD, DAC Online.)

destined for Latin America and the Caribbean also grew, but only by 22 per cent. As a result, the region's participation in global ODA dropped from more than 11 per cent in 2001 to 7 per cent in 2005 (Figure 11.1).

Flows of ODA to Latin America and the Caribbean have also fluctuated widely, especially in terms of bilateral donors. The principal sources of concessional lending and donations have been the United States, Japan, Germany, Spain and the European Commission, in that order. The main providers among multilateral organisations are the Inter-American Development Bank (IDB) (accounting for more than 50 per cent of the regional portfolios) and the Andean Development Corporation (ADC), which represents more than 50 per cent of the portfolios of Andean Community member countries. The World Bank still maintains an important position but has lost relative weight in the volume of lending, which has been partially compensated for by the regional banks. The Andean Community is the group of countries that received the largest volume of bilateral cooperation up to 2003, followed by the Center American Common Market (CACM) and Mexico.[4]

The boundaries of trade-related assistance (TRA) are hard to define, while the difficulties in measuring the magnitude of TRA abound (OECD/WTO 2006).[5] Nonetheless, according to the joint OECD/WTO report (2006), between 2002 and 2005 total AfT commitments from multilateral and bilateral donors rose by 22 per cent in real terms (at 2005 constant prices) and they committed on average US$ 21 billion per year for aid categories more closely related with aid for trade (AfT) (see Table 11.1). This represents a welcome contrast to the long-term declining trends since the mid-1970s. Between 2002 and 2005, the average share of AfT in total aid fluctuated between 32 per cent and 35 per cent. The AfT included on average US$ 11.2 billion to build economic infrastructure, US$ 8.9 billion to promote productive

TABLE 11.1. Aid for Trade, by category: (bilateral and multilateral), 2002–2005 (US$ millions at 2005 constant prices and by percentage)

	2002	2003	2004	2005
Trade policy & regulations	**817 (1.6)**	**614 (1.0)**	**478 (0.7)**	**654 (1.0)**
Economic infrastructure	**9,421 (18.5)**	**9,735 (16.2)**	**13,855 (19.8)**	**12,174 (18.0)**
Transport & storage	4,955	4,922	6,600	7,325
Communications	318	563	848	441
Energy	4,148	4,230	6,407	4,408
Productive capacity building	**7,587 (14.9)**	**9,388 (15.6)**	**9,759 (13.9)**	**8,918 (13.2)**
Banking & financial services	1,277	1,332	1,210	1,263
Business & other services	1,126	1,421	1,818	1,056
Agriculture, forestry, fishing	3,788	4,623	4,172	4,457
Industry, mining & construction	1,402	1,921	2,496	1,977
Tourism	45	91	63	164
Total Aid for Trade	**17,826 (34.9)**	**19,738 (32.8)**	**24,092 (34.4)**	**21,745 (32.2)**
Sector-allocable ODA	**51,051**	**60,232**	**69,973**	**67,528**

Source: WTO/OECD Aid for Trade at a Glance 2007: Ist Global Review, Table A1.1. Annex I, p.65.
Note: Figures in brackets correspond to a percentage of sector-allocable ODA.

capacities (including US$ 2 billion for trade development) and US$ 0.6 billion for increased understanding and implementation of trade policy and regulations.

During the period in question, the category of trade policy and regulations accounted for only 1.1 per cent of the total sector-allocable ODA, in strong contrast to the much higher shares corresponding to the other two categories (economic infrastructure and productive capacity building). Given the large scale of typical infrastructure projects, aid to support the development of economic infrastructure (transport and storage, communications and energy – a proxy for the WTO Task Force category "trade-related infrastructure") represented 54 per cent of the overall volume of AfT.

The increase in flows during the same period has, however, been insufficient to reverse the declining trend of AfT as a share of total sector-allocable ODA. The major factors behind this decline can be explained by the fall in public ownership of utilities in many OECD countries and a concomitant expansion of public-private partnerships.

In the framework of the AfT agenda, the increase in AfT is expected to be at least in the same order of ODA growth. In fact, Japan, the European Union (EU) and the United States announced a substantial increase in their AfT in the coming years.[6]

By recipient region, Asian countries receive on average more than twice the AfT destined for Africa. Recent statistics from the WTO/OECD database show that Latin America and Oceania received the smallest portion of resources[7] during 2002–2005. Indeed, the share of Latin America and the Caribbean in the global AfT total declined to 6.6 per cent in 2005, from an already low level of 9.1% in 2002 (see Table 11.2). The shares by category indicate that the region has been relatively well-placed in the category of productive capacity building, accounting for close to 11 per cent of the world total, while the share of economic infrastructure has stood only at 3.8 per cent.

TABLE 11.2. Latin America's share in Aid for Trade, 2002–2005 (US$ millions and percentages)

Latin America and the Caribbean	2002		2003		2004		2005		Baseline average 2002–2005 (volume, share)	
Trade policy & regulations	31	(3.8)	36	(5.9)	84	(17.6)	75	(11.5)	57	(8.9)
Economic infrastructure	503	(5.3)	309	(3.2)	346	(2.5)	539	(4.4)	424	(3.8)
Productive capacity building	1,086	(14.3)	870	(9.3)	1,083	(11.1)	830	(9.3)	967	(10.8)
Sub-total	1,620	(9.1)	1,215	(6.2)	1,513	(6.3)	1,444	(6.6)	1,448	(6.9)

Source: WTO/OECD Aid for Trade at a Glance 2007: Ist Global Review, Table A1.4. Annex I, p.68.
Note: Figures in brackets correspond to the percentage of total global Aid for Trade.

As a matter of fact, the list of the 25 largest recipients of aid for trade between 2002 to 2005 includes only one low middle-income Latin American country (Bolivia); the rest are mostly Asian and African countries including several Asian export "powerhouses" such as China, India and Thailand. Nevertheless, it is noteworthy that the position of Latin America and the Caribbean would certainly improve if the financial resources from regional organisations (not registered in the database) such as the Andean Development Corporation (ADC), the Financial Fund for the Development of the River Plate Basin (FONPLATA) and the Central American Bank of Economic Integration (CABEI) were taken into account. The financing from the Inter-American Development Bank fluctuated widely from year to year, passing for example from over US$ 2 billion in 2002 to just US$ 1 billion in 2004.

In a nutshell, although there has been a large and growing volume of resources committed for ODA and AfT in recent years, the share of Latin America and the Caribbean as a region has declined rapidly. This does not necessarily mean that the lack of financing had been the only determining factor behind the poorer-than-expected results. In fact, AfT stakeholders agree that, while resources do need to be increased, management must also be improved. It seems that the assignation of additional resources is a necessary, but not a sufficient condition, in order to translate trade liberalisation and the reforms of the multilateral trade system into welfare for the developing regions.

11.2.2 Management Disorder

The management problem is recognised by the Task Force on AfT. In its recommendations, the Task Force underlines the persistence of problems related to management, as summarised briefly in box 11.2. The concerns address the strategic and operational management of assistance, including the identification and preparation of projects, and not only the volume of financial resources and/or its inadequate distribution.[8] The responsibility of these problems is shared among beneficiaries, donors and intermediaries.

Box 11.1. Overall trends and major features of Aid for Trade

The joint WTO/OECD database, which includes non-Development Assistance Committee (DAC) donors and therefore differs substantially from the estimates made for the joint study from the same entities, reveals several interesting aspects of the overall use of aid for trade. The major points to be highlighted are:

- Assistance for *trade policy and regulations* (TPR) grew by more than 8 per cent since 2001–2002, reaching around US$ 900 million in 2005. This category includes activities that support the building of local capacities to: (i) formulate a broad national trade strategy; (ii) participate in trade negotiations; and (iii) implement trade agreements. *Trade facilitation* stands out among the subcategories, representing 37% of commitments in 2004. Other representative subcategories are *trade incorporation* in national strategies, the assistance for *technical barriers to trade and the negotiation and implementation of multilateral and regional agreements.*
- Support for *trade development* (which is part of *productive capacity building*) grew by almost 26 per cent, reaching US$ 2.173 billion in 2005. The most important subcategories were *trade promotion and market development*, which accounted for 28 per cent of *trade development* aid. The significant increase of assistance for e-commerce should be underlined, although it still represents a relatively low level.
- Contributions to *trust funds* for trade capacity building (TCB) (including the Joint Integrated Technical Assistance Programme (JITAP), the Integrated Framework (IF), the International Trade Centre (ITC) and WTO) grew by 17 per cent in 2005 with respect to 2003–2004, reaching US$ 55 million in 2005. During the four years under consideration, the total contribution reached over US$ 177 million, with strong support from the cooperation of several European countries (Denmark, Switzerland, Norway, the Netherlands and Sweden) and Canada.
- More than US$ 12 billion were annually destined to infrastructure assistance (not all directly linked to trade), with Asia being the principal beneficiary region.

Source: OECD (2006).

In general, technical assistance lacks a common framework and organisations do not usually know what their neighbour countries are doing, acting on a case-by-case basis and respecting mostly the donors' visions. There are additional restrictions because the nationality of the source of the resources is required to be the same in many cases. AfT is based on annual programmes but needs often transcend such a short time frame, triggering financing problems.

Box 11.2. Problems of trade-related technical assistance

√ Little attention paid to trade as a development instrument in beneficiary countries as well as in donor agencies.
√ Insufficient incorporation of trade in strategies for development and poverty reduction.
√ Scarce participation of the private sector in the identification of trade-linked needs.
√ Limited absorption capacity in beneficiary countries.
√ Inadequate mechanisms of associability and limited predictability in the donors' response to the priorities identified at the national and regional levels.
√ Lack of coordination and coherence in the donors' responses to trade-linked subjects.
√ Slow, overlapping and bureaucratic processes for trade technical assistance evaluation and supply, including excessive parallel structures in beneficiary countries.
√ Lack of information and analysis of trade policies and their impacts on development, scarce information available on AfT instruments.
√ Limited efficiency of the monitoring of trade-linked national policies and of donors' activities, absence of rigorous and independent evaluation of the results and impacts of projects and programmes.
√ Limited support for trade-linked regional, sub regional and cross-border programmes and projects.
√ Inadequate support to recognise and identify the costs of the adjustment associated with trade liberalisation.
√ Insufficient resources to create productive capacity and infrastructure.
√ Uneven coverage of country support.

Source: WTO (June 2006): Recommendations of the Task Force on Aid for Trade (WT/AFT/1).

Latin America and the Caribbean is not immune to this disorder. The problems identified herein apply, to a large extent, to the countries of the region, which consists of middle income economies with sometimes complex and sophisticated social and economic structures already in place. The existing mechanisms in the region share the following problems of AfT management.

- The absence of internal appropriation of the assistance programmes (due to the existence of multiple and conflicting agendas) has left TA/TCB highly dependent on donors' views and objectives, which do not necessarily coincide with those of the beneficiaries.

- Partly as a result of the experience of past decades, there are several organisations and agencies (official, private and NGOs) that "compete" in the ODA and TA/TCB markets, relegating the beneficiaries and their interests/needs to a position of secondary importance.
- In a framework of a complex institutional and political environment, competition in ODA and TA/TBA markets often translates into the fragmentation of aid and overlapping agendas and technical assistance projects, thereby endangering horizontal cooperation among public organisations and between these agencies and the private sector actors.
- Another important issue is the "proliferation" of regional trade agreements (RTAs), which results in multiple negotiation agendas and trade rules. As the World Bank states in its Global Economic Prospects 2005, "many RTAs have more life on paper than in reality" and "well designed agreements are of limited value if they are not implemented". RTAs should be considered by any AfT initiative, to the extent that the implementation and administration aspects of regional and bilateral FTAs (as well as the resulting "adjustment costs") are fully and effectively addressed.

Therefore, the viability of "additional, predictable, sustainable and efficient"[9] financing to reach the goal of AfT is linked to the creation of more effective and efficient management mechanisms. In order to truly translate trade into growth and welfare for beneficiary countries; there should be more efficient administration of resources allocated by donors and financers.

There exist important programmes managed by international organisations and donors such as the IF, JITAP, etc. The IF has addressed and resolved several problems and is now in a stage of redefinition. However, it is designed for LDCs, and it faces coordination problems. The proposals to improve its operation involve a significant increase of resources from USD 35 million to USD 400 million and, according to revised reports, not only would financial resources be increased, but its bureaucratic structures would also be improved.

At the Latin American level, TA/TCB has taken advantage of programmes offered by IDB, other organisations like the European Union, and the programmes and projects executed by the Andean Community. It is also relevant to underscore the Hemispheric Cooperation Program (HCP) in the framework of the Free Trade Agreement of the Americas (FTAA). The three participating organisations (IDB, OAS and ECLAC) undertook important AfT work, although the negotiations have regrettably reached a stalemate. Nonetheless, the HCP allowed the 34 member countries to develop national and subregional trade strategies that can be considered positive precedents for future AfT programmes at the national and regional levels. For instance, the AfT exercises carried out in the HCP framework were useful for the negotiation of bilateral trade agreements between Andean and Central American countries and the United States.

Any proposals should take into account these experiments in order to create an agile, practical and effective mechanism, which ought to complement the efforts already being made, and at the same time tackle the remaining problems and address specific regional demands and interests. The idea is to empower initiatives that overcome financing and management problems and would result in improved efficiency. Instead of overlapping efforts, new mechanisms should generate communication channels and programmes in the direction laid down in the Paris Declaration.

One of the prerequisites for the aforementioned approach is active *private participation* in project identification and development, financing and execution, as well as enhanced diagnostic and assistance services that can truly identify the peculiarities and AfT needs and interests of middle-income countries. This approach also calls for projects on trade infrastructure and project-finance possibilities through lending operations.[10] The major objective is to link up and complement ongoing efforts, so as to allow a more efficient channelling of resources.

In this regard, increasing the number of organisations or bureaucratic apparatus should be avoided. What is required is management, coordination, improved project evaluation processes, control and performance, as well as transparency and information exchange mechanisms, for the efficient use of present and future resources. If measures to improve management are not contemplated, just an increase in resources and higher expectations on the part of recipients are likely to further compound the problems and lead to the eventual dissatisfaction of donors and beneficiaries alike.

11.3. UNDERSTANDING AID FOR TRADE IN LATIN AMERICA AND THE CARIBBEAN

Latin American and Caribbean economies have become increasingly open and trade dependent; the share of trade in regional GDP has nearly doubled as a result of trade reforms, although it still remains below the levels of other developing regions. However, the most recent experience of the region shows that growth in exports of goods and services, is a necessary but not sufficient condition to promote economic development with social equity. Trade expansion should thus bring more support for sustainable development, the creation of high-quality employment and poverty reduction.

Latin American and Caribbean governments have been actively involved in negotiating a network of bilateral agreements for preferential liberalisation with countries both within the region and beyond. These agreements have generated centripetal and centrifugal forces that tend respectively to unify and fragment the process of regional integration. Some of the main concerns about these agreements relate to their complex nature and depth of their commitments, which could go beyond those agreed in the context of the World Trade Organisation (WTO). In addition, the rules and commitments assumed in the four regional integration schemes are considered to be much "shallower" than those in bilateral FTAs with the "Northern" industrialised countries. In this respect, there is an urgent need to establish a process of mutual complementarity and strengthening between the three roads of liberalisation and regulation.

The countries of the region have to assume high logistical costs on account of the weakness of their transport infrastructure, the lower level of development of logistics and transport services, and their slow and costly bureaucratic procedures for handling exports and imports. In this sense, the region urgently needs to establish clear measures and rules to facilitate business and trade activity, including the modernisation of customs procedures and harmonisation and compliance with international standards. In order to enhance international competitiveness, it is important to provide ECLAC member countries with policy analysis and technical cooperation on trade facilitation and other regulatory measures to be applied at the interregional, subregional, bilateral and national levels.

11.3.1 Factors Calling for More Aid for Trade at the National Level

11.3.1.1 Mainstreaming trade as key ingredient of national development strategies
Latin American and Caribbean governments increasingly recognise the importance of trade and competitiveness to their development objectives, and of making trade growth a key pillar of their national development strategies: "Supporting governments to further mainstream trade into the economic development agenda is one of the most pressing tasks of the-aid-for-trade initiative and a critical factor of its success" (IDB, 2007).

To achieve this goal, governments must involve an increasing number of stakeholders in devising and implementing trade and broader economic policies, with a view to ensuring public ownership of the reform process. The more successful that the AfT initiative can be in its conceptualisation and implementation, the more consensus there will be concerning the potential benefits of reform for civil society as a whole. The AfT initiative provides Latin American and Caribbean governments with a unique opportunity to develop a broader and better coordinated approach to TA/TCB in order to improve their participation in the world economy.

Part of this important challenge is policy coordination and priority-setting among the many government ministries and agencies including finance, agriculture, labour, and economic planning, as well as trade. Effective coordination among different levels of governments also requires a clear division of responsibility and labour between regional, national and local stakeholders. Regional institutions can play a significant role in identifying projects and in coordinating donors and countries across the region, while regional initiatives cannot be designed or implemented in isolation from national and local policies. Regional programmes are a complement, not a substitute, for national programmes (IDB/WTO, 2007).

The implementation of trade agreements is a case in point. This calls for revision of national development strategy, which includes trade policy, technological and productive development policies, social strategy and also macroeconomic policy. Moreover, changes in public and private institutions are needed, in order for them to adapt to the requirements of the new trade regime. At the national level, this requires (besides priority-setting), resources and capacity to create change-effective coordination between the different levels of government (central, regional, municipal) and between the former, civil society and the private sector.

All this goes far beyond mere legislation and changes to trade rules, since it can involve transcendental decisions in the area of productive policies, State modernisation, institutional development and infrastructure creation. With respect to employment and poverty, what is required is more than "compensation" for those affected by the cost of adjustment. Structural and technological changes, which generate more and better productive employment, are necessary.

11.3.1.2 Pressing needs to address supply-side constraints Despite the fact that some success has been achieved in the region in terms of trade diversification, the competitiveness of exports still depends mostly on price differentials arising from static comparative advantages, with very few but significant exceptions in the case of some countries. One pending task is to orient the trade agenda of the region not only towards strengthening access to markets for its labour- and natural-resource-intensive products and services, but also to increase the share of technology-intensive sectors.

The governments of Latin America and the Caribbean increasingly recognise the need to tackle supply-side problems to upgrade the quality of participation in the global markets. For this reason, there have been persistent and increasing needs for TA/TCB activities aimed at overcoming restrictions on supply, strengthening export capacity and "quality" through trade diversification by means of products with greater value added, knowledge content, compatible with international standards and markets requirements and promoting productive links between export sectors and the rest of the economy, especially small and medium-sized enterprises (SMEs), making use of the opportunities and advantages offered by the new Information and Communication Technologies (ICTs).

Despite a general consensus on the importance of supply-side constraints, opinions differ as to the orientation, scope and modalities of the assistance required. ECLAC suggests that AfT (at least for Latin American and Caribbean countries) should focus on three aspects: (i) strengthening national and regional institutions responsible for applying development-oriented trade policies and regulations; (ii) fostering private-sector initiative and entrepreneurship, and helping firms to become more competitive, with support from governments; and (iii) overcoming domestic constraints (trade facilitation and physical, human and institutional infrastructure).

11.3.1.3 Trade negotiation, implementation and administration at various levels Highly conscious of the benefits of free trade, establishing better and fair access to global markets, together with a rules-based international trading system, has become a major priority for the region's trade agenda, yet most countries still lack the necessary capabilities to negotiate and implement trade agreements. They happen to be currently involved in simultaneous negotiations, including the WTO global agenda, and other regional and sub-regional trade agreements and bilateral free trade agreements, with countries that are not necessarily in their geographical proximity.

The proliferation of FTAs in the region continues. ECLAC estimates suggest that approximately 60 per cent of Latin American and Caribbean exports in 2005/2006 were covered by preferential trade areas (PTAs) (i.e., bilaterals as well as plutilaterals) in one way or another, and that this proportion has shot up since the mid-1990s and continues to climb in the present decade. During this period, the most marked progress has been seen in the conclusion of FTAs with countries outside the region itself, especially with "Northern" countries such as the United States, Canada, the European Union (EU), the European Free Trade Association (EFTA) and Japan. In this process, Chile and Mexico have become true "semi-hubs" for FTAs in the Western Hemisphere.

This heralds a new phase in Latin American and Caribbean trade policy, as well as a definite reconfiguration of trading patterns that poses a formidable challenge for existing integration schemes. This, in turn, means that there is still a large portion of regional trade not covered by any trade preference. An increasing number of the countries in the region face the difficult task of not only negotiating more FTAs but also of simultaneously implementing and administrating those FTAs already signed. As the countries in the region move rapidly towards the implementation and adjustment stages of more comprehensive trade agreements, the scope of past technical assistance focused mainly on short-term negotiating needs related, may prove too narrow to ensure an efficient and equitable outcome of trade reform.

Another key aspect of trade liberalisation lies in reaching consensus between private sector representatives on complex – and often conflicting – issues resulting from various negotiations. Trade agreements can entail tangible benefits (for instance, exports growth and diversification) and intangibles ones (such as improved political and legal certainty) that are conducive to economic growth and improved social welfare. However, if there is an internal atmosphere of distrust among various sectors, and if national actors lack information about external trade opportunities and obligations and/or high implementation or adjustment costs, it is very possible that the agreements will not achieve the expected results. This requires a shift from a "top-down" consultation process to a "bottom-up" one, so that political impulses play a facilitating role in establishing a process that is validated by civil society and private sector actors.

For this reason, the growing demand for transparency on the part of stakeholders is changing the way trade negotiations are conducted and the processes through which trade policies are formulated, adopted and implemented. In this scenario, it is important for policymakers of the region to ensure that trade policies and practices -and the processes through which they are formulated and implemented- are both open and predictable. Transparency therefore becomes a fundamental requirement for social cohesion, helping to legitimate the agreements and facilitate their implementation.

In summary, adequate implementation requires an ability to generate policies, strategies and consensus to strengthen trade institutions, integrate new sectors and investors in the productive process, train the human resources needed by the markets and strengthen trade infrastructure. None of this will automatically result from trade liberalisation, even in its deepest form.

11.3.1.4 AfT as a promoter of public and private partnerships (PPP) Undoubtedly, although the AfT initiative ultimately benefits society as a whole, the private sector is the most immediately and directly affected by the costs and benefits deriving from trade reform and subsequent adjustment processes. As the trade agenda expands and deepens, more firms will be exposed to import competition and will benefit from greater market access opportunities provided by FTAs. Therefore, it is critical that the AfT initiative be designed in close collaboration with private-sector agents, with their active participation in the conceptualisation and implementation process from the outset. In this regard, the internationalisation of SMEs in the global market, associability and connectivity among these firms themselves and networking with Transnational Corporations (TNCs), as well as more enhanced insertion of Latin American and Caribbean SMEs in the world value chains, should receive high priority in the AfT agenda.

Furthermore, the private sector should become not only a recipient but also an active donor of AfT programmes. Latin American and Caribbean governments increasingly recognise the value of the private sector's technical expertise and knowledge base, and understand the importance of effectively engaging private-sector actors as partners both in designing and delivering AfT. Those companies with established experience in global supply-chain management can help other firms and even governments in harmonizing, standardizing and improving customs services, for example, with a view to facilitating customs procedures, international shipping and compliance with technical standards and sanitary and phytosanitary measures and trade facilitation.

Recent experience shows that private-sector involvement has not progressed much beyond support for trade negotiations, where defensive short-term interests have tended to prevail over long-term development goals. In this sense, the AfT initiative should focus on the latter by aiming to enhance international competitiveness based on the endogenous productive capabilities of national firms. However, for the private sector to stay engaged in the AfT initiative, it is imperative for tangible results to be delivered in a reasonable time frame. Because the time horizons of the private-sector are often short and its focus is more result-oriented, AfT programmes and projects need to be designed and delivered in a "business-friendly" manner.

11.3.1.5 Distributional and social-equity concerns Trade expansion should bring more support for sustainable development, the creation of high-quality employment and the reduction of poverty. During trade reforms in the region, efficiency considerations overshadowed the concerns on the distributional impact of trade liberalisation. As a consequence, governments did not consistently promote broad adjustment programmes to address disruption to local industries and labour markets caused by such reform: "Such disruptions, along with perceptions of insufficient public consultations, have tested LAC countries' continued support for trade liberalisation. In some countries, reform fatigue has heightened the risk of a backlash to more liberalisation." (IDB, 2007). The short-term and often painful process of reform, together with a longer period of time necessary for the realisation of many AfT projects, call for regular consultation between the government, private sector, labour and other social actors to sustain the momentum for trade reform.

In order to continue with, and further deepen, the reform process, it is necessary to adopt a comprehensive development strategy that makes trade work for the majority of the population. In summary, attention should be focused on the effects of trade liberalisation on the domestic economy, in particular from the perspective of economic and social equity and environmental protection. The AfT initiative should be conceived as an important component of this strategy.

11.3.1.6 Trade facilitation Trade facilitation has gained a special relevance in multilateral, regional and bilateral negotiations, because transaction costs associated with cumbersome, slow and unpredictable import and export processes can be higher than market access benefits obtained in trade agreements.[11] In an increasingly globalised economy, where tariff barriers are being rapidly reduced, trade facilitation measures are fast becoming a key area of international competitiveness.

In addition, efficient implementation of these measures is considered to have a strong and durable impact on FDI attraction because today's manufacturing sector is highly dependent on rapid, transparent and predictable border procedures. Besides, those countries that fail to modernise their border procedures risk being left out of international supply chains. Efficient border procedures also help to attract FDI in production areas that are labour-intensive, such as textiles and apparel and assembly-type activities in electrics and electronics.

Some of the constraints faced by developing countries cannot be solved nationally because they involve cross-border, sub-regional or regional issues. For example, investments in cross-border infrastructure are required to establish an efficient regional trade corridor, to facilitate cross-border transit or to manage the use of an asset common to several countries such as a river running through different countries.

The countries of the region have to assume high logistical costs on account of the weakness of their transport infrastructure, the less developed logistics and transport services, and their slow and costly bureaucratic procedures for handling exports and imports. The trade logistics-related indicators developed by the World Bank and the World Economic Forum, for example, confirm that in these aspects of trade competitiveness, Latin American and Caribbean countries substantially lag behind developed countries or developing countries of similar income-levels.

Trade facilitation mainly implies the simplification and rationalisation of the formalities, procedures, information and documents used in international trade, for the purposes of reducing transaction costs and raising competitiveness. Some central aspects to take into consideration include the following:

- Adaptation of laws, procedures and national documents to international agreements;
- Redesign of processes to eliminate trade obstacles, without sacrificing controls or pertinent regulations;
- Incorporation of information and communication technologies to optimise information flows; and
- Ensuring secure and safe transport, without sacrificing efficiency.

In the most recent regional agreements and the Doha negotiations, trade facilitation has assumed a primary importance. In the case of the Dominican Republic – Central America – United States Free Trade Agreement (CAFTA-DR), for instance, a separate chapter is dedicated to trade facilitation. The Doha agenda is based on three central aspects (transit, fees and formalities connected to trade, and publication and administration of trade regulations) and includes the mechanisms to ensure the application of the agreements (Special and Differential Treatment and technical assistance grants).

There is a strong rationality in countries' interest in establishing platforms to improve their trade procedures and regulations. At the multilateral level, the countries most advanced in this area before the culmination of negotiations will extract greater benefits from the Doha results. The signature of multiple free trade agreements, for its part, raises the possibility and the convenience of harmonizing rules and trade procedures between the various regional agreements, and as a result, the region as a whole can improve its competitiveness. The impact will be even greater for landlocked countries and their transit neighbours.

11.3.1.7 Resource-constraints faced by middle-income countries Latin America and the Caribbean is a region of mainly middle income countries but with high levels of debt. This both excludes them from donations and makes it difficult to generate fresh resources in financial markets. In this regard, one important priority for most countries is trade financing. The above-mentioned factors also highlight these countries' need for support, although this support should be different than the one given to the LDCs.

Many non-LDCs also need AfT support to mainstream trade into their national development strategies and ensure that relevant stakeholders such as the private sector and civil society actively participate in national consultations and the identification of challenges specific to the country. It has been suggested that a process similar to the one for LDCs should be set up and/or existing mechanisms improved for these countries. Given their middle-income status and their relatively mature trade institutions, the countries in the region need a specific set of AfT operative and monitoring instruments, including non-concessional lending and equity investment (IDB, 2007; IDB/WTO, 2007).

Although the WTO Task Force proposals on AfT tend to emphasise the needs of LDCs, only one country in Latin America and the Caribbean falls into this category.

ECLAC considers that the AfT analysis should take into consideration the special needs of Latin America and the Caribbean, in particular the landlocked countries and those at a lower stage of development. Therefore, any future work should rely on more flexible and case-by-case mechanisms and procedures in both funding and trade-related technical assistance.

11.3.1.8 Reform of trade-related institutions Institutions are key factors to determine transaction costs of trade and doing business, related to: (i) the efficiency in defining clear property rights and protecting them; and (ii) the efficiency of the enforcement and organisational arrangements. Institutions can be understood as a set of formal and informal rules conforming incentives, procedures and limits for organisations and people. Rules can be more or less efficient in terms of channelling resources to their most productive uses.

Latin American and Caribbean institutions are, in general, relatively well developed, as an offshoot of the "inward-looking development" period. However, existing institutions inherit many traits of bureaucracy and inefficiency, outdated for today's more open and globalised world economy. The change of rules (institutional changes) since the 1980s has been quite insufficient and series of reforms have been implemented unsatisfactorily or stayed trapped in the traditional mechanisms. Although the majority of the countries in the region have been undertaking structural adjustments, which include trade policy in a "multiple"[12] strategy framework, dating back to the beginning of the 1980s, the most common situation is the presence of advanced but incomplete or unsatisfactory reforms.

11.3.2 Aid for Trade for Regional Integration and Cooperation

Regional integration is both necessary and a matter of urgency. In addition to the traditional reasons for pursuing it, the current phase of the globalisation process generates integration-oriented demands, such as the need for strategic international alliances in the areas of production, logistics, marketing, investment and technology. The demand for competitiveness and technological innovation are mounting, while China's, India's and other Asian nations' competitive leapfrogging has redrawn the global map of trade flows and comparative advantages. Expanded markets, legal certainty and the convergence of rules and disciplines, in conjunction with advances in infrastructure, energy and connectivity, are now essential ingredients of growth with equity.

Unlike the situation in other regions, the Latin American regional integration process has been led by governments, not by market or businesses, and thereby tends to reproduce institutional problems at the regional level. Latin American and Caribbean regional integration has not been able to generate or articulate *de facto* with *de jure* integration, as currently seen in the Asia-Pacific region.

In this respect, AfT can support enhanced regional trade policy cooperation by playing an important role as a *catalytic instrument* to generate consensus and transparency between countries and regions involved in cross-border and subregional issues. Despite several setbacks in recent years, regional integration can, and should, foster a diversification of exports towards output that is more connected to the overall competitiveness of the economies concerned and that therefore helps to create dynamic comparative advantages and to give member countries options for tackling problems of asymmetry, income distribution and equity.

11.3.2.1 Harmonisation and convergence of norms and rules for "Deeper" regional integration There is much to be done if the Latin American and Caribbean region is to exploit fully the potential benefits of regional integration. The countries in the region should continue to work to overcome the constraints affecting their regional integration process. These constraints include: (i) perforations of common external tariffs and failure to bring customs union arrangements to completion; (ii) the persistence of non-tariff barriers; (iii) inadequate regional infrastructure; (iv) lack of effective subregional community institutions in several integration schemes; (v) limited coordination of macroeconomic and sectoral policies necessary for market integration and external trade and investment promotion; (vi) weak dispute-settlement mechanisms; and (vii) insufficient mechanisms for promoting a form of socio-economic development that would compensate for asymmetries in the distribution of the benefits of integration. The AfT initiative should support the ongoing efforts to converge rules and norms within and among the different customs unions.

Regional integration should tackle several dimensions of "deep integration" in a context of "open regionalism" by way of enhancing the provision of regional public goods (RPGs) (Kuwayama, 2005). This involves reducing tariff dispersion and non-tariff barriers, and addressing "behind-the-border" measures, while seeking to converge, if not harmonise, regulatory regimes, improve infrastructure (transport, communications, cooperation in energy via regional interconnection and ports), and strengthen dispute-settlement mechanisms.

Harmonisation or convergence of rules among member countries in areas such as services, investment, intellectual property rights, rules of origin, anti-dumping, safeguards, sanitary and phytosanitary norms, customs procedures, and factor mobility should enhance the systemic competitiveness of each country and the region as a whole. In addition, this approach to integration requires the adoption of agreements that will contribute to macroeconomic stability and productive development in each country. Government provision of many of these public goods is a key determinant of the enhancement of regional systemic competitiveness and several AfT initiatives should include measures to deepen regional integration.

Substantive differences exist between the scope of the rules and disciplines included in each type or category of agreement. Intraregional agreements tend to consist primarily of trade-remedy instruments designed to expedite tariff reduction schedules and avert the introduction of non-tariff trade barriers. These arrangements do not, however, provide for broader coverage of other types of rules which, because they are not fully harmonised, also act as non-tariff barriers (e.g., sanitary and phytosanitary measures and technical regulations). In addition, in the case of disciplines on trade-related matters (services, investment, government procurement, intellectual property), intraregional agreements fall far short of the coverage afforded by agreements with outside countries.

The coverage of commitments on trade disciplines also varies from one intraregional agreement to the next. South America's customs union arrangements and the associated bilateral agreements offer a high degree of coverage in terms of trade rules and dispute settlement, but this is not the case for sanitary and phytosanitary measures or technical barriers to trade. In contrast, the Free Trade Agreements (FTAs) signed between Chile and Mexico and by them with Central American countries provide more comprehensive coverage in these areas.

There is a striking asymmetry between bilateral and plurilateral agreements within the region and the agreements reached with countries elsewhere, especially some "Northern" countries. The latter contain more stringent commitments backed up by more binding mechanisms that entail greater legal certainty than subregional arrangements, which, by the same token, are less demanding in terms of trade and non-trade disciplines and rules.

The present assembly of bilateral, plurilateral and regional trade agreements in Latin America and the Caribbean could result in discrimination against some countries or subregional groups as a consequence of the wide variety of provisions in force with regard to coverage, difference in treatment and the depth of the commitments involved in the various disciplines and rules included in these agreements. Unless urgent steps are taken to achieve convergence among these different agreements, trade diversion will increase and the transactions costs for intraregional trade will climb.

From the perspective of Latin America and the Caribbean, four types of problems need to be addressed: (i) operational issues, such as customs, transit and storage procedures and rules, where a lack of familiarity with formalities or confusion about methods of application can become hidden trade barriers; (ii) the presence of rules and disciplines that are formalised in some agreements but absent from others (investment, services and intellectual property, for example), that have depth and coverage in intraregional agreements that differ from agreements with outside countries, or that have differing provisions regarding similar issues (national treatment) or treat identical subjects differently (commitment or negotiation models); (iii) the institutional structure of trade agreements, and (iv) discrimination between trading partners (less favourable treatment) as a result of differences between the regulations, policies and liberalisation measures of intraregional and extra regional agreements'.

Convergence can be promoted by various sorts of stimuli and different sorts of modalities, which should be examined on a topic-by-topic basis. In order to do so, a flexible outlook that is conducive to creative solutions must be maintained.

While recognizing countries' differing sizes and trade orientations, the gains of existing integration schemes must be preserved by fostering convergence on trade and non-trade issues alike. Certainly, the countries belonging to each individual integration scheme must ask themselves what that scheme is doing to contribute to their growth and competitiveness. Be that as it may, prevailing conditions make it advisable to place a higher priority on regional cooperation than on trade negotiations per se.

Current integration efforts should focus on establishing common ground as a platform for convergence in energy and infrastructure policies, first of all, and, subsequently in policies on the environment, tourism, connectivity, ICTs, e-commerce, regulatory practices and other matters. If regional cooperation efforts in these areas succeed in rebuilding confidence among the countries, then, in addition to paving the way for competitiveness gains, they will make it less difficult to then build bridges among the various intraregional trade arrangements by defining a basic set of shared obligations and flexible timetables, particularly for the smaller economies, together with infrastructure, trade facilitation and connectivity programmes that would provide for special and differential treatment.

11.3.2.2 AfT as a driver of de facto regional integration and south–south cooperation Another aspect affecting Latin America's international integration in the global market is its regional trade integration, *de facto* less developed than in Asia and East and Central Europe. Regional integration offers many advantages that support world integration: regional trade agreements allow integration with similar neighbours, increase the scale of production and sales and intra-industry trade, promote the creation of international supply chain networks, constitute a learning platform to compete in other more demanding markets, regional blocs attract more FDI and have greater bargaining power than individual countries.

In the American continent, there is an intense intraregional and intra-industrial trade between the United States and Mexico and, to a lesser extent, with Central America and some countries of the Caribbean. North and Central America strengthened their trade relations with the NAFTA in 1994 and then more recently with the CAFTA-DR. In South America, regional trade is far less significant and is in search of a new dynamic. The two main integration systems, the Andean Community and MERCOSUR, are involved in a process of redefinition after the Bolivarian Republic of Venezuela left the former and is in the process of becoming a member of the latter.

11.3.2.3 Physical infrastructure and sustainable development In Latin America and the Caribbean, there are many issues and major challenges in the field of infrastructure[13] and sustainable development that have motivated regional cooperation. Transport and tourism are representative examples. The issue of transport has gained importance in the debate on instruments to promote trade and investment in a world of low tariffs and progressive elimination or harmonisation of non-tariff barriers, since lower transport costs directly encourage exports and imports and are equivalent in effect to a tariff reduction.

At the same time, market failures and imperfections that raise transport costs tend to concentrate industrial and economic activity in areas that are already endowed with suitable human and physical infrastructure. Some major initiatives include the South American Regional Infrastructure Integration Initiative (IIRSA) and the Puebla-Panama Plan (PPP). The Association of Eastern Caribbean States has also set up a programme entitled "Unifying the Caribbean by Air and Sea", which aims to harness public and private efforts among member countries, and stimulate regional cooperation mechanisms. Caribbean countries will also cooperate in implementing air transport training programmes.

In terms of sustainable development, the Latin American and Caribbean region has a characteristic that distinguishes it from all others, both in terms of the wealth and importance of its natural resources and in the global risks implied by the rapid process of environmental degradation. In recent years, several major regional proposals have emerged in this domain. The eight signatory countries of the PPP adopted the Meso-American Sustainable Development Initiative as a strategic and cross-cutting framework to ensure that all relevant projects, programmes and measures contain appropriate environmental management practices and promote conservation and sustainable management of natural resources.

Another important example is the Regional Biodiversity Strategy for the Tropical Andean Countries: a wide-ranging initiative for this subregion that represents one of the world's richest zones in terms of natural resources, containing about 25% of the biological diversity of the entire planet. The strategy is one of the first attempts by the subregion to develop a comprehensive platform of community action, promoting cooperation between member countries and projecting a new and unique image to the international community. It is also one of the first community strategies adopted on this issue by a group of signatory countries of the Convention on Biological Diversity, and makes a specific contribution towards fulfilling its objectives.

11.3.2.4 Social cohesion and structural funds Another important area for AfT action regarding regional integration relates to social cohesion. Despite its recognised importance, the social dimension has not been adequately tackled by the Latin American and Caribbean regional integration schemes. The implementation of the social agenda is usually slow or postponed, while the concern for distributive effects has been taken up only sporadically. The mechanisms of regional and subregional agreements provide only limited financial resources aimed at improving cohesion among social groups and supporting regions. Therefore, one of the major challenges in Latin American regional integration has been the adoption of effective, sustainable economic development policies and social integration policies for all social sectors.

In effect, integration processes have been significantly influenced by the following realities: (i) disparities among the States and regions pose a risk for the integration process itself; (ii) the market alone does not promote economic integration when

States and regions display such disparities; (iii) growth, competitiveness and employment are the principal instruments for achieving social cohesion; (iv) there is a strong need to integrate economic and social policies in order to guarantee the citizenry a universal social protection system; (v) the reduction of disparities by way of better access to training, education, employment (of women, young people and the long-term unemployed) as well as closing income and wealth gaps as an important base for social cohesion; and (vi) strong political will to support States and less developed regions is necessary in order to achieve the conditions of convergence.

11.3.2.5 Building on past experience of AfT Conscious of the lack of capacity in implementing all the rules and norms contemplated in bilaterally with developed countries, Latin America and the Caribbean included several provisions on TA/TCB in their FTAs with the "Northern" countries. The association agreements signed by Chile and Mexico with the European Union, for example, include a series of cooperation activities directed toward facilitating implementation. At the regional level, the parties to the Free Trade Area of the Americas (FTAA) agreed to set up a Hemispheric Cooperation Program (HCP), helping countries determine their needs and help donors target their TA. In a similar vein, in the United States-Central American and United States-Andean negotiations, the parties developed National Action Plans, targeting three areas: (i) preparation for negotiations; (ii) implementation of resulting agreements, and (iii) adjusting to integration. These efforts were considered valuable since the parties were often excluded from similar multilateral initiatives such as the IF.

Although the AfT initiative can build on past efforts, it would have to translate needs into specific TA programmes. The initiative is a welcome outcome since the countries in the region may be reluctant to re-engage in time-consuming needs-assessment exercises without assured funding.

Despite the fact that the countries in the region have not participated in previous multilateral initiatives such as the IF, they have accumulated experience in linking aid with trade. This should enable the region to move quickly from diagnosis to action. This time, however, the task is more difficult: instead of focusing excessively on short-term negotiation needs, the TA/TCB activities should aim at boosting international competitiveness and an efficient and equitable insertion into the world economy, a process involving the interaction of many complex long-term factors.

11.3.3 Recommendations of the Aid-for-Trade regional review for Latin America and the Caribbean[14]

This meeting was organised by the Inter-American Development Bank (IDB) and WTO, and was hosted by the government of Peru, in cooperation with the World Bank. For the first time, it brought together trade and finance ministers, bilateral donors, regional and multilateral agencies and private-sector representatives to discuss how to build the capacity and infrastructure that countries need to take advantage of today's dynamic global economy. An important objective of the meeting was to shift the focus of the Aid-for-Trade initiative from policy debate to implementation, and to promote more effective policies. It was also considered as a regional exercise, prior to the WTO Global Aid-for-Trade Review held in Geneva, 20–21 November 2007.

The meeting concluded that the following four areas should be of high priority:

1. Trade-related infrastructure and regional public goods
2. Trade finance and SME export development
3. Trade facilitation and standards
4. Export promotion and export diversification.

To address these objectives, IDB and WTO put forward the following recommendations for consideration by the Global Review on Aid for Trade meeting, November, 2007 in Geneva.

- Encourage countries and subregions to map out their main structural constraints and policy bottlenecks to trade, building on existing studies and needs assessments;
- Update national and regional action or plans to address key priorities, including the identification of timetables, institutional mandates, and financial resources for implementation;
- Endorse the constitution of an " Aid-for Trade LAC Support Network" – led by IDB and WTO, including countries, regional secretariats, lead donors, multilateral and regional agencies and private-sector representatives – to assist countries and subregions in mapping out priorities and developing action plans;
- Explore funding for priority pilot projects with existing available resources and/or request the creation of specific regional financial facilities to fund projects included in the action plans; and
- Prepare a road map and timetable for monitoring progress, including a regional "stock-taking" conference to be held in the region to be followed by an opeara-tionalisation conference.

To mobilise AfT for the region, public-private partnerships were considered essential. The meeting emphasised the importance of finding creative ways to engage private-sector actors in designing and implementing trade strategies. AfT should be mobilised to build private-private institutions that can partner governments on policy development and also on implementation of TATCB activities and projects.

The participants emphasised that certainty on the scale of the AfT initiative would be crucial to secure sustained commitment in the region. As implementation of the initiative begins, the participants considered that concrete actions would be needed in order to meet the most pressing challenges and to ensure that existing momentum would not dissipate.

Given the limited access of the countries in the region to ODA grant resources and concessional lending, in the view of the participants, non-concessional lending and equity investment would be key in addressing the region's TATCB needs. In this regard, AfT grants can provide crucial seed money for large infrastructure programs and other supply-side problems which would require large non-concessional financing. In their view, a key issue was how multilateral lending could be made more accessible, for example, by facilitating "blending" with donor assistance. Another important issue was how to increase the incentives for private investment in TCB, and to expand the scope for public and private partnerships.

The participants considered that there is potential for large returns from regional investments, as coordination failures may jeopardise the optimal provision of regional public goods (RPBs), and that the availability of grant financing for RPBs may help attenuate this problem. Also, the complexity of aligning national sovereign guarantees to access to lending facilities for regional projects might call for the development of new regional financial instruments.

11.3.4 The ECLAC Aid-for-Trade Agenda and possible contribution to the initiative

In the background note to the High-level Committee on Programmes (HLCP) of the United Nations System Chief Executives Board for Coordination (CEB), the WTO Secretariat calls for the support of United Nations agencies and other international organisations like OECD, IMF, the World Bank, regional development banks and other relevant international agencies. WTO defines coherence as critical to AfT and as a test of WTO ability to work more cooperatively. The same paper states also that the challenge is not to invent a new mechanism, but rather to get the existing mechanisms to work together more effectively.

In Latin America and the Caribbean, WTO has a close partnership with the IDB and, within this framework; partnerships exist with other regional organisms and institutions.[15] ECLAC can cooperate in several ways, considering that one of its main institutional objectives is to strengthen insertion in the global economy and regional integration of Latin American and Caribbean countries.

Under its general programme of work in the field of trade and development, ECLAC carries out a subprogramme on International Trade and Integration. It places increasing emphasis on the development of local capacity, especially in areas related to:

- Implementation and administration of trade agreements (including regional integration efforts), which requires countries to review, reform or even upgrade their policies, strategies and institutional framework;
- Monitoring and support on trade negotiations (at various levels – multi, pluri, regional and bilateral) to assure equitable distribution of benefits and costs;
- Identification of supply-side constraints that impair trade diversification by product, destination, sector and firm size, and the promotion of greater incorporation

of value added and technology and knowledge contents into exports, not only in primary products but also manufactures;
- Analysis of the role of innovation as a facilitator of international competitiveness;
- Identification and implementation of appropriate trade facilitation measures (technological upgrade of customs procedures, sanitary and phytosanitary measures, technical barriers to trade, rules of origin, and physical infrastructure including maritime transport and ports) and AfT instruments at the country level;
- Search for convergence of norms and rules within and among subregional integration schemes and provision of Regional Public Goods (RPGs) to promote regional integration;
- Diffusion of information on trade related matters, focused on emerging markets such as China, India and ASEAN countries and best practices on trade and investment promotion; and
- Building of networks and public and private sector linkages.

In this way, present and future work programmes will complement and support the WTO AfT initiative.

The tasks undertaken by ECLAC in *support of administration and implementation* functions, arising from the signing of preferential and free trade agreements, start from the basis that the exploitation of those agreements requires countries to review, reform or even upgrade, their policies, strategies and institutional framework, especially in relation to issues only partially covered with by WTO ("WTO-plus" issues). In accordance with emerging views, ECLAC has also been proposing *the coordination of rules in various domains —origin, standards and trade facilitation, among others—* in addition to regional-scale investment and financing efforts that would stimulate and facilitate trade between countries.

The conceptualisation process of AfT for Latin America and the Caribbean also involves a regional dimension: *regional integration as a vehicle for the enhancement of international competitiveness and trade diversification*. There is an urgent need to address the question of how to: (i) stimulate regional trade flows; (ii) reform regional integration institutions to face the challenge of the multiplication of bilateral trade agreements with countries within the region and beyond; and (iii) proceed with the convergence of trade-related rules and regulations at subregional levels that is necessary for such trade reactivation. AfT measures also include investments and South-South trade expansion and technology transfer, and they call for a more important role for regional banks in supporting trade capacity.

ECLAC has received many requests for assistance from member countries in the AfT sphere, and has successful organised many training courses and seminars. To carry out the above mentioned tasks effectively and efficiently, as a United Nations regional commission, ECLAC is well positioned to assess real needs of countries at the national and regional levels. ECLAC can put to use its renowned *expertise to undertake diagnostic or pre-feasibility studies*. For AfT conception and implementation, the role of ECLAC lies in information gathering on and monitoring mechanisms for country needs, while *it also has the appropriate institutional framework for the management, coordination and provision of aid at the country or regional level.*

Together with institutional and physical infrastructure to strengthen supply-side capacity and competitiveness, ECLAC is currently executing several joint projects on the above-mentioned TCB areas with agencies such as the German Agency for Technical Cooperation (GTZ), the Andean Development Corporation and IDB, as well as regional integration organisations such as the Latin American Free Trade Association (ALADI), MERCOSUR, the Andean Community, the Central American Common Market (CACM) and the Caribbean Community (CARICOM). In the last two years, *ECLAC has intensified joint activities with WTO, especially in the area of trade negotiation/administration, while it has actively participated in training courses offered by both WTO and INTAL/IDB.*

As the previous joint efforts with other regional agencies testify, the extensive experience of ECLAC in regional development processes complements the expertise that IDB offers in AfT. The bank's program of work points to the urgency of comprehensive strategies that make trade useful to development. This complementarity is evident especially in the area of trade negotiation/implementation: the coexistence of different kinds of preferential agreements (bilateral, plurilateral, regional and even inter-regional) with integration schemes of different depths adds to the complexity of the AfT agenda and thus to the need of trade capacity building.

There is a fertile ground for collaboration between IDB and ECLAC in the area of regional integration, especially from the perspective of regional public goods (RPBs). Apart from the AfT agenda itself, infrastructure including Information and Communications Technologies (ICTs), energy and social cohesion are important emerging issues for the countries in the region; ECLAC has initiated several projects to make concrete progress in this respect. The two agencies can pool resources to strengthen countries' trade-related capacities by creating cooperation forums or networks among the countries in the region and projects linking trade facilitation and investment in order to contribute to the participation of SMEs in the international arena and infrastructure initiatives (IIRSA and PPP). In short, the regional perspective seems to be gaining ground in terms of both analyses and AfT initiatives. As a result, there is a greater potential role to be played by regional organisations.

Furthermore, there is also room for collaboration between the two institutions in the analysis of the distributional aspects of trade, since both are committed to assisting the countries of the region in utilizing trade as a driver of growth and poverty reduction. There are important synergies to be derived between trade-related development assistance and the experience of ECLAC in providing assistance for capacity building to member countries. More specifically, it would be important to collaborate on assisting countries in developing and adopting pro-poor trade-related strategies and complementary policies that would enable them to increase high-quality employment and improve the incomes of the poor. ECLAC can contribute to research, simulating dialogue and launching specific initiatives on trade and poverty at the national and regional levels, helping the region to make trade work for the majority of the population.

Given its mission and experience, ECLAC could collaborate with IDB and WTO on the AfT initiative with three complementary initiatives:

- Through the identification of *specific projects and initiatives* to empower the integration between trade and sustainable development, including private sector organisations and propitiating regional integration;
- Through its participation in the *design and implementation of regional coordination and cooperation mechanisms* so as to implement the AfT agenda at the regional level; and
- Supporting the WTO monitoring and evaluation work from the regional perspective, participating in evaluations and in the consultative group suggested by this organisation.

More specifically, ECLAC is well placed to support the AfT initiative by working with governments and regional integration organisations, in close collaboration with various private-sector stakeholders, and supporting:

- National efforts for the implementation and administration of bilateral and multilateral trade agreements in subjects linked to organisational development, critical subjects (such as dispute settlement), productive supply problems (working with the affected productive sectors) and evaluation of the impacts of the agreements.
- Definition of action plans in the trade facilitation area so as to reduce trade-related transaction costs, including small and landlocked countries, in the framework WTO trade facilitation group.
- Formation and functioning of **AfT National Committees** recommended by the Task Force to be organised in aid receptor countries in order to strengthen their national coordination, complement the action of other organisations and serve as a mechanism to articulate the AfT agenda. Their members would be governments, private sector and donors, under the direction of the government's corresponding ministry.

In parallel to these activities, ECLAC can put in place regional coordination and cooperation mechanisms:

- To develop a regular mechanism (for instance, a Trade Forum similar to the one proposed by the Economic and Social Commission for Asia and the Pacific (ESCAP)) to share AfT experiences between countries of the region, donors, international agencies and the private sector, in close cooperation with other regional organisations
- To identify and implement technical assistance programs to support the AfT agenda, with support from WTO and other regional organisations, taking stock

of the experiences of the IF and the HCP (Tripartite Committee) as a reference. These exercises will be guided by the following principles; (i) **coherence and coordination** to ensure the efficient use of resources from the perspective of both beneficiaries and stakeholders, and (ii) **concrete and verifiable results** from the perspective of beneficiaries and other actors.

These initiatives should contribute to make the national and regional AfT agenda more articulated and productive, complementing, not substituting or duplicating the work of other national or international agencies.

In addition, ECLAC can contribute to the monitoring and evaluation of AfT initiatives and projects, by:

- Providing support for the information compilation and systematisation for the joint WTO/OECD database, coordinating and/or supervising the process at the regional level;
- Complementing and supporting the WTO *Trade Policy Reviews* giving basic information on the countries in the region, as well as preparing reports, which could serve as internal materials for WTO; the TPRs of the countries in the region should be probably prepared with a higher frequency, in the framework of AfT agenda.
- Serving as host, organiser and catalyzing agent for the ***Aid-for Trade LAC Support Network*** suggested by the report of the regional meeting of September 2007 (IDB/WTO 2007). This mechanism could be similar to the Regional Committee proposed by the AfT Task Force, integrating regional and sub regional financial organisations and institutions in charge of supervising the AfT regional aspects and presenting reviews of the regional and sub regional needs, on the basis of the Task Force recommendations.

ECLAC can assume these challenges considering that one of its main institutional objectives is *to strengthen insertion in the global economy and regional integration of Latin American and Caribbean countries*. ECLAC also has certain expertise in this area: ECLAC played an important role as a member of the *Tripartite Committee* (with IDB and OAS) created to coordinate activities of the Hemispheric Cooperation Program (HCP) in the FTAA (although this process is currently suspended). The HCP developed methodologies to assist governments with carrying out needs-assessments to identify the many challenges faced during the initial steps of AfT process.[16]

In any region, effective AfT solutions require a "systemic" and "interdisciplinary" approach that takes into consideration a whole range of transmission channels of trade effects to economic and social development; it is of extreme importance to address trade links with other areas such as sustainable development, social development, infrastructure including transport systems, natural-resource endowment, and entrepreneurship and technological progress. In this respect, the comprehensive understanding that ECLAC has of the region's development process and the close contacts it maintains with individual countries and other regional organisations would enable it to provide significant support to the region's AfT undertaking. Furthermore, ECLAC is equipped with substantive divisions with great expertise in each of above-mentioned areas, and is therefore well suited to assume a leadership role in conceptualizing, operationalizing and implementing the AfT Initiative for Latin American and Caribbean countries.

NOTES

1 WTO (December, 2005): Hong Kong ministerial declaration. WT/MIN(05)/3/Rev.2.

2 Following the recommendations of the Paris Declaration on Development Aid Efficiency (2 March 2005), an international agreement signed by over 100 Ministers, heads of agencies and other senior officials who committed their countries and organisations to continue to increase efforts in harmonisation, alignment and management of aid, with a view to achieving results with a set of monitorable actions and indicators.

3 OECD Database (DAC Online, 2007).

4 Latin American Economic System (SELA): "La ayuda oficial para el desarrollo en América Latina y el Caribe, Caracas, October, 2005.

5 The Task Force specified six categories of AfT: (i) trade policy and regulations; (ii) trade development, covering investment promotion, analysis and institutional support for trade in services, public-private sector partnership, e-commerce, trade finance, trade promotion, market analysis and development; (iii) trade-related infrastructure; (iv) productive capacity building, covering support to address trade-related supply-side constraints; (v) trade-related adjustments, to help countries put in place accompanying measures that allow the countries to exploit benefits of trade liberalisation; (vi) other trade-related needs, not covered in the preceding categories. Trade development was later included in productive capacity building.

6 WTO (July, 2006).

7 WTO (May, 2006). Task Force on AfT, communication from Colombia, Ecuador and Peru.

8 In fact, both aspects can be considered management problems.

9 WTO (July, 2006).

10 According to the Doha working chronogram for 2006, the proposal to improve the Integrated Framework included several of these improvements as topics for discussion.

11 Using the data on time it takes to transport cargo from the factory gate to the ship in 126 countries, Djankov et al. (2006) find that each additional day that a product is delayed prior to being shipped reduces trade by more than 1%, which is equivalent to a country distancing itself from its trade partners by about 70 kilometres on average. They conclude that reducing export times by 10 days is likely to have a bigger impact on exports of developing countries than any feasible liberalisation in Europe or North America. Similarly, Hummels (2007) finds that tariff equivalents for export delays exceed tariffs faced by exporters in developing countries.

12 See Devlin and Vodusek (2005)

13 The region needs to spend "more" and "better" on infrastructure. Countries in the region spend , on average, less than 2% of GDP on infrastructure, whereas 3% to 6% is needed to keep pace with other countries like China or the Republic of Korea (González et al. 2007)

14 Lima, Peru, 13-14 September, 2007. For the complete report, see "Mobilizing Aid for Trade: Focus Latin-America and the Caribbean. Report and Recommendations", prepared by the Inter-American Development Bank (IDB) and the World Trade Organisation (WTO), November 2007.

15 WTO (2004).

16 The HCP gave technical support to 29 of the 32 developing countries participating in FTAA negotiations. Several country strategy papers were prepared (included trade-related capacity-building and training needs) and donor roundtables were organised to allow donors to identify and coordinate their support.

REFERENCES

Devlin R. and Z. Vodusek (2005). Trade related capacity building: an overview in the context of LA trade policy and the MERCOSUR-EU association agreement, INTAL/IADB Occasional Paper No. 29, Washington, D.C., March.

Djankov, S., C. Freund and C.S. Pham (2006). Trading on time, World Bank Policy Research Working Paper No. 3909, May 2006.

González, J.A., J. L. Guasch and T. Serebrisky (2007), Latin America: Addressing High Logistics Costs and Poor Infrastructure for Merchandise Transportation and Trade Facilitation, Washington: The World Bank.

Hummels, D. (2007), Calculating tariff equivalents for time in trade, USAID Report, March.

IDB (2006). Aid for Trade: The Inter American Development Bank's Experience in Latina America and the Caribbean, Washington, D.C., Integration and Regional Programs Department, May.

IDB (2007). Mobilizing Aid for Trade: Latin America and the Caribbean. Discussion Paper, Washington: IDB.

IDB/WTO (2007). Mobilizing Aid for Trade in Latin America and the Caribbean. Draft Discussion Paper, Washington, D.C., September.

Kuwayama, M. (2005). Latin American South-South integration and cooperation: from regional public goods perspective, Multilateralism and Regionalism: The New Interface, UNCTAD/DITC/TNCD/2004/7, Geneva, United Nations Conference on Trade and Development (UNCTAD).

OECD (2005). Paris Declaration on Aid Effectiveness, Paris, High Level Forum, 28 February- 2 March.

OECD/WTO (2006). 2006 Joint WTO/OECD Report on Trade-Related Technical Assistance and Capacity Building, Paris: OECD/WTO.

OECD/WTO (2007). Aid for Trade at a Glance 2007: 1st Global Review, Geneva: OECD/WTO.

World Bank (2005). Global Economic Prospects, Washington: World Bank.

WTO (2004). Technical Assistance and Training Plan, WT/COMTD/W/133/Rev.2, Geneva, December.

WTO (2005). Doha Work Programme. Ministerial Declaration, WT/MIN(05)/DEC, Geneva, December.

WTO (2006), Recommendations of the Task Force on Aid for Trade, WT/AFT/1, Geneva, July.

STATISTICAL ANNEX

ANNEX I Selected data on regional arrangements[1]

Regional arrangements	Year of establishment	Website(s)	Regional share of world gross domestic product at current prices in US dollars (2004)[2]	Regional share of the world's total population (2005)[3]	Intra-regional trade as a percentage share of the region's total trade (2005)[4]	Intra-regional trade intensity index (2005)[5]
African Union (AU)	2002	http://www.africa-union.org/	1.79	13.45	7.01	11.3
African, Caribbean and Pacific Group of States (ACP)	1975	http://www.acp.int/	1.51	12.32	8.9	14.45
Andean community (CAN)	1969	http://www.comunidadandina.org/	0.49	1.49	10.62	22.73
Arab Maghreb Union	1989	http://www.maghrebarabe.org/	0.44	1.21	1.42	13.92
Asia-Pacific Economic Co-operation (APEC)	1989	http://www.apec.org/	54.33	40.97	70.03	1.43
Association of Caribbean States (ACS)	1994	http://www.acs-aec.org/	2.62	3.95	6.77	2.22
Association of Southeast Asian Nations (ASEAN)	1967	http://www.aseansec.org/	1.88	8.61	24.8	4.37

(Continued)

ANNEX I (*Continued*)

Regional arrangements	Year of establishment	Website(s)	Regional share of world gross domestic product at current prices in US dollars (2004)[2]	Regional share of the world's total population (2005)[3]	Intra-regional trade as a percentage share of the region's total trade (2005)[4]	Intra-regional trade intensity index (2005)[5]
Baltic Sea States Subregional Co-operation (BSSSC)	1993	http://www.bsssc.com/	4.66	3.31	22.22	3.4
Bay of Bengal Initiative for Multi-sectoral Technical and Economic Co-operation (BIMSTEC)	1997	http://www.bimstec.org/	2.28	21.11	3.57	1.29
Benelux Customs Union (BENELUX)	1944	http://www.benelux.be/	2.17	0.42	14.94	2.16
Black Sea Economic Co-operation (BSEC)	1992	http://www.bsec-organization.org/	3.15	5.07	16.77	4.11
Caribbean Community (CARICOM)	1973	http://www.caricom.org/	0.11	0.35	9.33	81.57
Central American Integration System	1991	http://www.sgsica.org/	0.2	0.51	6.27	54.89

Central European Free Trade Agreement (CEFTA)	1992		0.14	0.28	9.23	34.71
Common Market for Eastern and Southern Africa (COMESA)	1994	http://www.comesa.int/	0.51	6.63	2.95	15.78
Common Market of the South (Mercosur)	1991	http://www.mercosur.int/	2.16	4.06	14.53	8.08
Community of Sahel-Saharan States (CEN-SAD)	1998	http://www.cen-sad.org/	0.81	6.68	4.19	11.7
Council of Arab Economic Unity (CAEU)	1957	http://www.caeu.org.eg/	0.64	3.14	3.38	19.99
Council of Europe (COE)	1949	http://www.coe.int/	34.92	12.27	73.9	1.58
Council of the Baltic States (CBSS)	1992	http://www.cbss.st/	11.42	4.6	22.95	1.42
Cross Border Initiative	1992		0.18	3.08	7.15	64.33

(Continued)

ANNEX I (*Continued*)

Regional arrangements	Year of establishment	Website(s)	Regional share of world gross domestic product at current prices in US dollars (2004)[2]	Regional share of the world's total population (2005)[3]	Intra-regional trade as a percentage share of the region's total trade (2005)[4]	Intra-regional trade intensity index (2005)[5]
East African Community (EAC)	1999	http://www.eac.int/	0.08	1.56	7.14	431.36
Economic and Monetary Community of Central Africa (CEMAC)	1994		0.09	0.58	1.23	49.18
Economic Community of Central African States (ECCAS)	1983	http://www.ceeac-eccas.org/	0.16	1.98	0.71	56.21
Economic Community of the Great Lakes Countries (ECGLC-CEPGL)	1976		0.08	0.36	0.03	0.18
Economic Community of West African States (ECOWAS)	1975	http://www.ecowas.int/	0.29	4.02	6.25	102.54

Economic Co-operation Organisation (ECO)	1977	http://www.ecosecretariat.org/	1.58	6.16	5.47	2.55
European Free Trade Association (EFTA)	1960	http://www.efta.int/	1.53	0.19	0.62	0.3
European Union	1957	http://www.europa.eu.int/	31.71	7.61	68.54	1.61
Eurozone[6]	1999	http://www.ecb.int/	23.35	4.85	51.06	1.63
Greater Arab Free Trade Area (GAFTA)[7]	2005		1.76	4.13	5	3.37
Gulf Cooperation Council (GCC)	1981	http://www.gcc-sg.org/	1.09	0.56	2.31	2.21
Indian Ocean Commission (IOC)	1984	http://www.coi-info.org/	5.06	1.29	0.81	0.17
Indian Ocean Rim Association for Regional Cooperation (IOR-ARC)	1997		6.25	28.86	24.35	2.74

(Continued)

ANNEX I (*Continued*)

Regional arrangements	Year of establishment	Website(s)	Regional share of world gross domestic product at current prices in US dollars (2004)[2]	Regional share of the world's total population (2005)[3]	Intra-regional trade as a percentage share of the region's total trade (2005)[4]	Intra-regional trade intensity index (2005)[5]
Intergovernmental Authority on Development (IGAD)	1996	http://www.igad.org/	0.13	2.95	3.04	72.99
Latin American and Caribbean Economic System (SELA)	1975	http://www.sela.org/	4.99	8.62	17.81	3.55
Latin American Integration Association (LAIA)	1980	http://www.aladi.org/	4.63	7.64	16.3	3.26
League of Arab States	1945	http://www.arableagueonline.org/	2.04	4.84	4.81	3.52
Mano River Union (MRU)	1973	http://www.manoriverunion.org/	0.01	0.28	–	–
Mekong River Commission (MRC)	1995	http://www.mrcmekong.org/	0.52	2.62	2.35	2.52
Melanesian Spearhead Group (MSG)	1993		0.03	0.12	0.36	101.12

Nordic Council and the Nordic Council of Ministers	1952	http://www.norden.org/	2.55	0.38	20.45	5.69
North American Free Trade Agreement (NAFTA)	1994	http://www.nafta-sec-alena.org/	32.89	6.79	46.9	2.3
Organisation of Eastern Caribbean States (OECS)	1981	http://www.oecs.org/	0.01	0.01	1.21	565.42
Organisation of Islamic Conference (OIC)	1969	http://www.oic-oci.org/	5.02	21.96	9.17	1.89
Organization of American States (OAS)	1948	http://www.oas.org/	36.22	13.76	52.8	2.3
Pacific Islands Forum	1971		1.83	0.50	9.23	6.84
South Asian Association for Regional Co-operation (SAARC)	1985	http://www.saarc-sec.org/	2.12	22.57	4.2	2.39
Southern African Customs Union (SACU)	1910	http://www.sacu.int/	0.57	0.84	3.35	6.28

(Continued)

ANNEX I *(Continued)*

Regional arrangements	Year of establishment	Website(s)	Regional share of world gross domestic product at current prices in US dollars (2004)[2]	Regional share of the world's total population (2005)[3]	Intra-regional trade as a percentage share of the region's total trade (2005)[4]	Intra-regional trade intensity index (2005)[5]
Southern African Development Community (SADC)	1980	http://www.sadc.int/	0.72	3.56	8.32	16.39
West African Economic and Monetary Union (WAEMU)	1994	http://www.uemoa.int/	0.25	3.17	4.87	202.12
Western European Union (WEU)	1954	http://www.weu.int/	26.95	5.51	53.37	1.58

[1] See Annex II for membership information.
[2] Calculated by RIKS© based on UN Statistics Division: GDP at current prices – US dollars series.
[3] Calculated by RIKS© based on UN National Accounts Main Aggregate Database, Population series.
[4] Calculated by RIKS© based on UN COMTRADE Database, SITC Rev.1.
[5] Intra-regional trade intensity index is defined as the ratio of the intra-regional trade share to the region's share in the world's total trade.
[6] The indicators are based on data of the 11 member states that initiated the Eurozone in 1999 and Greece, that was admitted on 1 January 2001.
[7] The calculations of the indicators do not include data of Palestine.

ANNEX II Membership of regional arrangements, as of January 1, 2008

African Union	Algeria
	Angola
	Benin
	Botswana
	Burkina Faso
	Burundi
	Cameroon
	Cape Verde
	Central African Republic
	Chad
	Comoros
	Congo
	Cote d'Ivoire
	Democratic Republico f the Congo
	Djibouti
	Egypt
	Equatorial Guinea
	Eritrea
	Ethipia
	Gabon
	Gambia
	Ghana
	Guinea
	Guinea-Bissau
	Kenya
	Lesotho
	Liberia
	Libya
	Madagascar
	Malawi
	Mali
	Mauritania
	Mauritius
	Mozambique
	Namibia
	Níger
	Nigeria
	Rwanda
	Sahrawi Arab Democratic Republic
	Sao Tome and Principe
	Senegal
	Seychelles
	Sierra Leone
	Somalia
	South Africa
	Sudan
	Swaziland
	Tanzania

(*Continued*)

	Togo
	Tunisia
	Uganda
	Zambia
African, Caribbean and Pacific Group of States	Angola
	Antigua and Barbuda
	Barbados
	Belize
	Benin
	Botswana
	Burkina Faso
	Burundi
	Cameroon
	Cape Verde
	Central African Republic
	Chad
	Comoros
	Congo
	Cook Islands
	Cote d'Ivoire
	Cuba
	Democratic Republic of Congo
	Djibouti
	Dominica
	Dominican Republic
	East Timor
	Equatorial Guinea
	Eritrea
	Ethiopia
	Federated States of Micronesia
	Fiji
	Gabon
	Gambia
	Ghana
	Grenada
	Guinea
	Guinea-Bissau
	Guyana
	Haiti
	Jamaica
	Kenya
	Kiribati
	Lesotho
	Liberia
	Madagascar
	Malawi
	Mali
	Marshall Islands
	Mauritania

	Mauritius
	Mozambique
	Namibia
	Nauru
	Niger
	Nigeria
	Niue
	Palau
	Papua New Guinea
	Rwanda
	Saint Lucia
	Saint Vincent and the Grenadines
	Samoa
	Sao Tome and Principe
	Senegal
	Seychelles
	Sierra Leone
	Solomon Islands
	Somalia
	South Africa St. Kitts and Nevis
	Sudan
	Suriname
	Swaziland
	Tanzania
	The Bahamas
	Togo
	Tonga
	Trinidad and Tobago
	Tuvalu
	Uganda
	Vanuatu
	Zambia
	Zimbabwe
Andean Community	Bolivia
	Colombia
	Ecuador
	Peru
Arab Maghreb Union	Algeria
	Libya
	Mauritania
	Morocco
	Tunisia
Asia-Pacific Economic Co-operation	Australia
	Brunei Darussalam,
	Canada
	Chile
	China
	Chinese HongKong
	Chinese Taipei
	Indonesia

(Continued)

	Japan
	Korea
	Malaysia
	Mexico
	New Zealand
	Papua New Guinea
	Peru
	Philippines
	Russia
	Singapore
	Thailand
	United States
	Vietnam
Association of Caribbean States	Antigua & Barbuda
	Araba
	Barbados
	Belice
	Colombia
	Costa Rica
	Cuba
	Dominica
	Dominican Republic
	El Salvador
	France
	Grenada
	Guatemala
	Guyana
	Haiti
	Honduras
	Jamaica
	Mexico
	Netherland Antilles
	Nicaragua
	Panama
	Saint Kitts Nevis
	Saint Lucia
	Saint Vincent and the Grenadines
	Suriname
	The Bahamas
	Trinidad & Tobago
	Turks and Caicos
	Venezuela
Association of Southeast Asian Nations	Brunei
	Cambodia
	Indonesia
	Laos
	Malaysia
	Myanmar
	Philippines

	Singapore
	Thailand
	Vietnam
Baltic Sea States Subregional Co-operation	Denmark
	Estonia
	Finland
	Germany
	Latvia
	Lithuania
	Norway
	Poland
	Russian Federation
	Sweden
Bay of Bengal Initiative for Multi-sectoral Technical and Economic Co-operation	Bangladesh
	Bhutan
	India
	Myanmar
	Nepal
	Sri Lanka
	Thailand
Benelux Customs Union	Belgium
	Luxembourg
	Netherlands
Black Sea Economic Co-operation	Albania
	Armenia
	Azerbaijan
	Bulgaria
	Georgia
	Hellenic Republic
	Moldova
	Romania
	Russian Federation
	Serbia
	Turkey
	Ukraine
Caribbean Community	Antigua & Barbuda
	The Bahamas
	Barbados
	Belize
	Dominica
	Grenada
	Guyana
	Jamaica
	Montserrat
	Saint Kitts and Nevis
	Saint Lucia
	Saint Vincent and the Grenadines
	Suriname
	Trinidad and Tobago

(*Continued*)

Central American Integration System	Belice
	Costa Rica
	El Salvador
	Guatemala
	Honduras
	Nicaragua
	Panama
Central European Free Trade Agreement	Albania
	Bosnia and Herzegovina
	Croatia
	Kosovo
	Macedonia
	Moldova
	Montenegro
	Serbia
Common Market for Eastern and Southern Africa	Burundi
	Comoros
	Democratic Republic of the Congo
	Djibouti
	Egypt
	Eritrea
	Ethiopia
	Kenya
	Libya
	Madagascar
	Malawi
	Mauritius
	Rwanda
	Seychelles
	Sudan
	Swaziland
	Uganda
	Zambia
	Zimbabwe
Common Market of the South	Argentina
	Brazil
	Paraguay
	Uruguay
	Venezuela
Community of Sahel-Saharan States	Benin
	Burkina Faso
	Chad
	Central African Republic
	Comoros
	Côte d'Ivoire
	Djibouti
	Egypt
	Eritrea
	Gambia
	Ghana
	Guinea

	Guinea-Bissau
	Liberia
	Libya
	Mali
	Morocco
	Niger
	Nigeria
	Sierra Leone
	Senegal
	Somalia
	Sudan
	Togo
	Tunisia
Council of Arab Economic Unity	Egypt
	Iraq
	Jordan
	Kuwait
	Libya
	Mauritania
	Palestine
	Somalia
	Sudan
	Syria
	Yemen
Council of Europe	Albania
	Andorra
	Armenia
	Austria
	Azerbaijan
	Belgium
	Bosnia and Herzegovina
	Bulgaria
	Croatia
	Cyprus
	Czech Republic
	Denmark
	Estonia
	Finland
	France
	Georgia
	Germany
	Greece
	Hungary
	Iceland
	Ireland
	Italy
	Latvia
	Liechtenstein
	Lithuania
	Luxembourg

(*Continued*)

	Macedonia
	Malta
	Moldova
	Monaco
	Montenegro
	Netherlands
	Norway
	Poland
	Portugal
	Romania
	Russian Federation
	San Marino
	Serbia
	Slovakia
	Slovenia
	Spain
	Sweden
	Switzerland
	Turkey
	Ukraine
	United Kingdom
Council of the Baltic States	Denmark
	Estonia
	Finland
	Germany
	Iceland
	Latvia
	Lithuania
	Norway
	Poland
	Russian Federation
	Sweden
Cross Border Initiative	Burundi
	Comoros
	Kenya
	Madagascar
	Malawi
	Mauritius
	Mozambique
	Namibia
	Rwanda
	Seychelles
	Swaziland
	Tanzania
	Uganda
	Zambia
	Zimbabwe
East African Community	Burundi
	Kenya

	Rwanda
	Tanzania
	Uganda
Economic and Monetary Community of Central Africa	Cameroon
	Central African Republic
	Chad
	Congo
	Equatorial Guinea
	Gabon
Economic Community of Central African States	Angola
	Burundi
	Cameroon
	Central African Republic
	Congo
	Democratic Republic of Congo
	Equatorial Guinea
	Gabon
	Sao Tome and Principe
Economic Community of the Great Lakes Countries	Burundi
	Democratic Republic of Congo
	Rwanda
Economic Community of West African States	Benin
	Burkina Faso
	Cape Verde
	Cote d'Ivoire
	Gambia
	Ghana
	Guinea-Conakry
	Guinea-Bissau
	Liberia
	Mali
	Níger
	Nigeria
	Senegal
	Sierra Leone
	Togo
Economic Co-operation Organization	Afganistán
	Azerbaijan
	Iran
	Kazakhstan
	Kyrgyzstan
	Pakistan
	Tajikistan
	Turkey
	Turkmenistán
	Uzbekistan
European Free Trade Association	Iceland
	Liechtenstein
	Norway
	Switzerland

(*Continued*)

European Union	Austria Belgium Bulgaria Cyprus Czech Republic Denmark Estonia Finland France Germany Greece Hungary Ireland Italy Latvia Lithuania Luxembourg Malta Netherlands Poland Portugal Romania Slovakia Slovenia Spain Sweden United Kingdom
Eurozone	Austria Belgium Cyprus Finland France Germany Greece Ireland Italy Luxembourg Malta Netherlands Slovenia Spain
Greater Arab Free Trade Area	Bahrain Egypt Iraq Jordan Kuwait Lebanon Libya Morocco Oman

	Palestine
	Qatar
	Saudi Arabia
	Sudan
	Syria
	Tunisia
	United Arab Emirates
	Yemen
Gulf Cooperation Council	Bahrain
	Kuwait
	Oman
	Qatar
	Saudi Arabia
	United Arab Emirates
Indian Ocean Commission	Comoros
	Madagascar
	Mauritius
	Réunion France
	Seychelles
Indian Ocean Rim Association for Regional Cooperation	Australia
	Bangladesh
	India
	Indonesia
	Iran
	Kenya
	Madagascar
	Malaysia
	Mauritius
	Mozambique
	Oman
	Seychelles
	Singapore
	South Africa
	Sri Lanka
	Tanzania
	Thailand
	United Arab Emirates
	Yemen
Intergovernmental Authority on Development	Djibouti
	Eritrea
	Etiopia
	Kenya
	Somalia
	Sudan
	Uganda
Latin American and Caribbean Economic System	Argentina
	Bahamas
	Barbados
	Belice
	Bolivia

(*Continued*)

	Brazil
	Chile
	Colombia
	Costa Rica
	Cuba
	Dominican Republic
	Ecuador
	El Salvador
	Grenada
	Guatemala
	Guyana
	Haiti
	Honduras
	Jamaica
	Mexico
	Nicaragua
	Panama
	Paraguay
	Peru
	Suriname
	Trinidad and Tobago
	Uruguay
	Venezuela
Latin American Integration Association	Argentina
	Bolivia
	Brazil
	Chile
	Colombia
	Cuba
	Ecuador
	Mexico
	Paraguay
	Peru
	Uruguay
	Venezuela
League of Arab States	Algeria
	Bahrain
	Comoros
	Djibouti
	Egypt
	Iraq
	Jordan
	Kuwait
	Lebanon
	Libya
	Mauritania
	Morocco
	Oman
	Palestine

	Qatar
	Saudi Arabia
	Somalia
	Sudan
	Syria
	Tunisia
	United Arab Emirates
	Yemen
Mano River Union	Guinea
	Liberia
	Sierra Leone
Mekong River Commission	Cambodia
	Laos
	Thailand
	Vietnam
Melanesian Spearhead Group	Fiji
	Papua New Guinea
	Solomon Islands
	Vanuatu
Nordic Council and the Nordic Council of Ministers	Denmark
	Finland
	Greenland
	Iceland
	Norway
	Sweden
North American Free Trade Agreement	Canada
	Mexico
	United States
Organization of Eastern Caribbean States	Anguilla
	Antigua and Barbuda
	British Virgin Islands
	Dominica
	Grenada
	Montserrat
	Saint Kitts and Nevis
	Saint Lucia
	Saint Vincent and the Grenadines
Organization of Islamic Conference	Albania
	Algeria
	Azerbaijan
	Bahrain
	Bangladesh
	Benin
	Cameroon
	Chad
	Comoros
	Cote d'Ivoire
	Djibouti
	Gabon

(*Continued*)

	Gambia
	Guinea
	Guinea-Bissau
	Guyana
	Indonesia
	Iraq
	Kazakhstan
	Kuwait
	Kyrgyz Republic
	Lebanon
	Malaysia
	Maldives
	Mali
	Morocco
	Mozambique
	Niger
	Oman
	Palestine
	Qatar
	Saudi Arabia
	Senegal
	Sierra Leone
	Somalia
	Sudan
	Suriname
	Syrian Arab Republic
	Tajikistan
	Togo
	Tunisia
	Turkey
	Turkmenistan
	Uganda
	United Arab Emirates
	Uzbekistan
	Yemen
Organization of American States	Antigua and Barbuda
	Argentina
	Bahamas
	Barbados
	Belice
	Bolivia
	Brazil
	Canada
	Chile
	Colombia
	Costa Rica
	Cuba
	Dominica
	Dominican Republic

	Ecuador
	El Salvador
	Grenada
	Guatemala
	Guyana
	Haiti
	Honduras
	Jamaica
	Mexico
	Nicaragua
	Panama
	Paraguay
	Peru
	Saint Kitts and Nevis
	Saint Lucia
	Saint Vincent and the Grenadines
	Suriname
	Trinidad and Tobago
	United States
	Uruguay
	Venezuela
Pacific Islands Forum	Australia
	Cook Islands
	Federated States of Micronesia
	Fiji
	Kiribati
	Marshall Islands
	Nauru
	New Zealand
	Niue
	Palau
	Papua New Guinea
	Samoa
	Solomon Islands
	Tonga
	Tuvalu
	Vanuatu
South Asian Association for Regional Co-operation	Afghanistan
	Bangladesh
	Bhutan
	India
	Maldives
	Nepal
	Pakistan
	Sri Lanka
Southern African Customs Union	Botswana
	Lesotho
	Namibia

(*Continued*)

	South Africa
	Swaziland
Southern African Development Community	Angola
	Botswana
	Democratic Republic of the Congo
	Lesotho
	Madagascar
	Malawi
	Mauritius
	Mozambique
	Namibia
	Seychelles
	South Africa
	Swaziland
	Tanzania
	Zambia
West African Economic and Monetary Union	Benin
	Burkina Faso
	Cote d'Ivoire
	Guinea-Bissau
	Mali
	Nigeria
	Senegal
	Togo
Western European Union	Belgium
	France
	Germany
	Greece
	Italy
	Luxembourg
	Netherlands
	Portugal
	Spain,
	United Kingdom

Printed in the United States
138839LV00001B/9/P

9 781402 094545